Marine Chemistry

The Author

Professor D. Satyanarayana is a distinguished Marine Chemist. He served in Andhra University (1967-97) in several academic positions as Head of the Department, Chairman, PO Board of Studies and Director, School of Chemistry. He initiated a full- fledged two year specialization of marine chemistry at postgraduate level, the first of its kind in the Indian Universities and taught over 20 years.

He has varied experience in research covering solvent extraction and coordination chemistry of lanthanides, Marine pollution, geochemistry of marine sediments, biogeochemical cycling of nutrients and trace metals in the marine environment. He trained 15 doctorate students and published over 100 papers in various National and International Journals.

Prof. Satyanarayana visited several countries–Germany, Canada, UK, France, Austria and participated in International seminars. He was awarded Dr. E.C. La Fond Medal for presenting best paper from the developing countries in IAPSO (International Association of Physical Sciences of Oceans) held at Vienna (1991). He was associated with several National and International committees–Indian Scientific committee on Ocean Research (SCOR), Research Council of National Institute of Oceanography, International Commission for Cooperation with Developing Countries. He is currently associated as member with Coastal Ocean Monitoring and Prediction Systems (COMAPS) of DOD, Government of India; Coastal Zone Management Authority (CZMA) and Shore Area Development Authority (SADA), Government of Andhra Pradesh; and as consultant of Delta Studies Institute, Andhra University.

Marine Chemistry

By

D. Satyanarayana

Former Professor of Marine Chemistry
School of Chemistry, Andhra University
Visakhapatnam – 530 003

2026

Daya Publishing House®

A Division of

Astral International Pvt. Ltd.

New Delhi – 110 002

First Published, 2010
Reprinted, 2026
ISBN 9788170354581

Disclaimer:

Every possible effort has been made to ensure that the information contained in this book is accurate at the time of going to press, and the publisher and author cannot accept responsibility for any errors or omissions, however caused. No responsibility for loss or damage occasioned to any person acting, or refraining from action, as a result of the material in this publication can be accepted by the editor, the publisher or the author. The Publisher is not associated with any product or vendor mentioned in the book. The contents of this work are intended to further general scientific research, understanding and discussion only. Readers should consult with a specialist where appropriate.

Every effort has been made to trace the owners of copyright material used in this book, if any. The author and the publisher will be grateful for any omission brought to their notice for acknowledgement in the future editions of the book.

Published by : **Daya Publishing House®**
A Division of
Astral International Pvt. Ltd.
– ISO 9001 : 2015 Certified Company –
4736/23, Ansari Road, Darya Ganj
New Delhi-110 002
Ph. 011-43549197, 23278134
E-mail: info@astralint.com
Website: www.astralint.com

Digitally Printed at : **Replika Press Pvt. Ltd.**

PRINTED IN INDIA

Acknowledgements

It is a pleasure to express appreciation to Mr. B.V. Sridhar Rao for diligently transferring the manuscript and diagrams into presentable form. The author gratefully acknowledge the following sources for illustrations used in this book.

Figures 1.1, 1.2, 1.4, 3.1, 3.2, 3.3, 3.9 and 4.2 from Garry Bearman (Ed., 1995) Sea water: Its composition, properties and behaviour Vol.2(2nd edn.) Pergamon/Open University Press; Figures 2.2, 4.10 and 4.11 from Garry Bearman (Ed., 1995), Ocean Chemistry and Deep Sea Sediments Vol.5, Pergamon/Open University Press; Figures 1.3, 2.1(A, B) and 2.3 from J.B. Wright (Ed., 1978) Chemical Processes, Units 7 and 8, Open University Press; Figures 7.1, 7.2 and 7.3 from J.B. Wright (Ed., 1978) Biological Environments, Unit 9, Open University Press; Figures 4.1, 4.3 and 4.4 from E. Olausson and E.Cato (Eds., 1960) Chemistry and Biogeochemistry of Estuaries, John Wiley and Sons; Figures 3.7, 3.8 and 4.5 from R.A. Horne (1969), Marine chemistry, Wiley Interscience; Figures 1.5, 3.5, 4.6 (a, b), 6.1, 6.2 and 7.4 from R. Sen Gupta and E. Desa (Eds., 2001). The Indian Ocean – A perspective, Vol. 1 and 2, Oxford and IBH (New Delhi); Figures 4.12, 4.13, 5.1, 5.2 and 5.3 from W.S. Broecker (1974) Chemical Oceanography, Hartcourt, Inc.; Figure 3.6 from Ittekote and R.R. Nair

(Eds., 1993) Monsoon Biogeochemistry, SCOP and UNEP, Hamburg; Figure 3.11 from B.N. Desai (Ed., 1992) Oceanography of the Indian Ocean, Oxford and IBH, and Figure 3.4 from K. Wyrtki (1988), Oceanographic Atlas of the International Indian Ocean Expedition, National Science foundation.

D. Satyanarayana

Foreword

History and culture in India have from ancient times, been closely woven with matters relating to the ocean. Despite this long linkage to the ocean, its formal scientific study remains an almost nascent subject in India. This is unfortunate, as almost no other discipline better exemplifies the convergence of the basic sciences as well as does Oceanography.

Of the various scientific disciplines traditionally associated with the study of oceanography, Marine Chemistry remains the most recent inclusion in the curricula of Indian Universities. This is unfortunate as Marine Chemistry plays a pivotal role in understanding the oceans – from its circulation patterns to biological productivity, from mineral formation to global geo-chemical cycles. A result of this neglect of marine chemistry is similarly reflected in the paucity of available text books on the fundamentals of aquatic chemistry, and further compounded in that none of the oceanographic books now being published by Indian authors focus on the fundamentals of the subject. The present book adequately satisfies the need not only of oceanographers but also of geoscientists and environmentalists in India. Prof. D. Satyanarayana must be lauded for undertaking this onerous but important task.

Prof. Satyanarayana brings to this work the vast experience and insight that he has gained during his teaching of ocean chemistry over the last 25 years at Andhra University, Visakhapatnam. His experience and deep knowledge of the subject has given his presentation lucidity where even the most difficult topics are explained in simple terms suitable for new entrants to the subject.

The topics have been largely organized along the lines of traditional chemical classification of elements in sea water, yet cover processes as varied as the acquisition of the chemical composition of ocean waters to global biogeochemical cycles – the latter being of current concern not just to environmentalists but to policy makers as well. Each chapter presents subject basics, global distributions and specific information on the Indian Ocean. The latter information will go a long way in ensuring that new entrants to Marine Chemistry research have a sound data set on our surrounding seas to begin their work with. It will also be of special use for students and researchers of other countries, particularly those surrounding the Indian Ocean.

Our global environment remains under threat from our changing life styles and our unreasonable demands on earth resources, and publicly funded research is increasingly and urgently addressing the issues of environmental sustainability and degradation. Prof. Satyanarayana' s work comes at a crucially important time by providing a cogent and formal text well suited both the students and environmental researchers, and in this process serves not only academia but society as well.

Ehrlich Desa
Former Director, NIO, Goa
Currently Head, C.B.S.,
I.O.C., UNESCO,
Paris–75015

Preface

The book intends to provide latest information on chemical processes in oceans and distribution of chemical constituents in world oceans with special emphasis on the Indian Ocean. It is designed in such a way that it can be read on its own or as part of marine science/ocean science and technology/fishery/earth/ environmental science courses at postgraduate level. The book can be easily followed by practicing scientists and engineers with basic knowledge in chemistry, working on marine environment and related areas.

Chapter 1 deals with the early developments in oceanography world wide and the factors that culminated the development on marine chemistry as an emerging discipline in recent years. The concept of salinity, its definition, methods of determination and spatial distribution in World Oceans with special reference to the Indian Ocean are discussed.

Chapter 2 introduces the chemistry of major and minor elements their average abundance in oceans, and characteristic features. It summarises the classification of elements into conservative, recycled and scavenged elements, their common vertical profiles in oxidising and reducing environments.

Chapter 3 describes dissolved gases, particularly oxygen and carbon dioxide, their exchange between air-sea interface and interaction with living and non-living matter in the sea. Involvement of carbon dioxide in chemical processes such as CO_2 – carbonate equilibria, its role as a buffer in controlling the pH and solubility of calcium carbonate in sea water are explained.

Chapter 4 begins with the biogeochemistry of nutrient elements (nitrogen, phosphorus and silicon) essential for sustenance of life in the sea. It includes nutrient budgets, analytical methods of determination and cycles in the sea. Vertical and lateral movements of nutrients are explained taking a typical example of phosphate, in terms of a simple two box model. Their stoichiometric relationships with oxygen are explained in terms of average atomic ratios (carbon: nitrogen: phosphorus) in sea water and phytoplankton.

Chapter 5 reviews the importance of radio-isotopes as tracers in the study of ocean processes such as vertical mixing, gas exchange, geochronology of sediments, and growth rates of manganese nodules in the deep ocean. It begins with the brief description and chemistry of radio nuclides, their classification, occurrence, concentration and activity levels in sea water.

Chapter 6 provides a brief description of organic matter in the sea, classification into dissolved and particulate fractions, sources, nature and removal processes and seasonal and spatial distributions in the sea. It includes interrelationships of organic matter and nutrients with primary production.

Chapter 7 discusses the basic principles of photosynthetic productivity considered as most important stage in the marine food chain. The chemical reactions and mechanism of this complex process leading to phytoplankton production, analytical methods for its measurement and factors controlling primary production in the sea are explained.

References cited in the text and a list of books for further reading are incorporated at the end of the book. Each chapter contains multiple choice, short answer and review questions to inculcate curiosity and to test one' s own understanding of the book.

Dr. Ehrlich Desa, former Director of National Institute of Oceanography (NIO) and currently working as Head, Capacity Building Section, Inter Governmental Oceanographic Commission

(IOC), UNESCO, Paris, has kindly agreed to write a foreword focusing the nature and scope of the book. I am indebted to Dr. Rabin Sen Gupta, Chief project coordinator, Gujarat Ecological Society for his critical comments and valuable suggestions for improving the quality of the text. I should particularly express my thanks to Dr. V.V. Sarma, Deputy Director, Regional Centre of NIO, for taking all the pains for reading the entire drafts and making helpful comments on them. Thanks are also due to Prof. Anil K. De, Department of Chemistry, Viswabharati, Santiniketan; Prof. G.S. Roonwal, Dept. of Geology, Delhi Unuversity, my former colleague, Prof. U. Muralikrishna and Research students, who encouraged me directly or indirectly for undertaking this work. I am thankful to Dr. K.S.R. Murthy, Deputy Director and Scientist-in-Charge, Regional Centre of NIO for providing the technical and secretarial assistance.

D. Satyanarayana
Former Professor of Marine Chemistry
Andhra University, Visakhapatnam (A.P)

Contents

Chapter-1
Introduction

1.1.0 Early Developments in Oceanography

Systematic oceanographic studies began from the Challenger Expedition (1872-1876) sponsored by the Royal Society of London. The multi disciplinary expedition covered a distance of around 68,000 nautical miles (1,25,336 km), collected 133 loads of rocks and sediments from the ocean floor and took 492 soundings. All ocean basins except the Arctic had been covered and 4,717 new animal species had been collected and classified. The expedition produced unprecedented mass of observations (covering the data in 50 volumes written by 76 authors over a period of 23 years) in physics, chemistry, biology and geology of oceans. This expedition is often regarded as the basic stage in the development of Oceanography. The Challenger Expedition was followed by several other important expeditions (1911 to 1940) undertaken by the countries such as Germany, U.S.A., Denmark, France, Italy, Norway, Netherlands and Russia, leading to systematic and dynamic nation-wide ocean surveys.

However, modern oceanography dates back to 1957 (Wust 1964). By then many scientists and governments realized that small, nationally based sea going programmes could no longer probably make major scientific contributions. Large scale international cooperation was clearly the way of the future. A major step in this direction was taken by the International Council of Scientific Unions (ICSU), which organized the International Geophysical Year (IGY) during 1957 to 1958 wherein scientists from 67 nations participated. Coordinated studies were conducted on tides, currents, earth heat flow, gravity, magnetic fields and ocean basin topography. These studies led to the sea floor spreading theory of Alfred Weginer. Similarly, the Scientific Committee on Ocean Research (SCOR) and Inter-governmental Oceanographic Commission (IOC) of UNESCO organized the International Indian Ocean Expedition (IIOE) involving 23 countries, 40 research vessels and 180 cruises (1959 to 1965). This largest multiship assault on the vast unknown (Indian) Ocean provided new information about physics, chemistry, biology and geology of the ocean (Wooster 1984), which was hitherto almost unknown.

In 1974, United Nations convened a conference on the Law of Sea (UNCLOS). The world nations considered oceans as the common heritage of mankind. The UNCLOS (1982) stipulated that (i) within 200 nautical mile (370 km.) zone, called the Exclusive Economic Zone (EEZ), coastal states exercise sovereignty with regard to the management of all living and non-living resources; beyond the EEZ, the resources are the common heritage of mankind; and (ii) coastal states must not exploit and endanger living resources. The Montreal Protocol and the subsequent conventions at London, and Copenhagen made provisions of technology transfer and financial support to developing countries (Strong, 1992). Thus enormous areas of oceans came under the jurisdiction of each nation. Within the EEZ, coastal states have rights and responsibilities equal to those they have on land and they can regulate the living (fisheries) and nonliving resources (minerals, oil, energy, etc.). After the required minimum ratification, this convention (UNCLOS III) came into effect from 14th November 1994. India ratified UNCLOS III in June 1995.

Studies on remote sensing of the ocean surface from air crafts and satellites dramatically improved the capability to map surface ocean currents and watermasses. Instruments on earth-orbiting satellites can measure ocean surface temperature, map surface waves and winds and detect colour changes caused by the growth of phytoplankton in shallow waters and by suspended sediments in river waters flowing into the sea. These surface and near surface measurements provide snap shots of ocean processes on the scale that cannot be obtained by observation from the research ships.

1.1.1 Developments of Oceanography in India

The first scientific investigation of the Indian Ocean was carried out by HMS Challenger (1873–74) in the South Indian Ocean. In early 1950' s ocean research in India was devoted mainly to fish distribution and fishery ecology, but some observations on seasonal variations of sea surface temperature, chemical composition of sea water and plankton were included in the local and isolated areas. The Indian Meteorological Department (IMD) engaged in collection of data on sea state, swell, waves and currents. Oceanographic studies in the Bay of Bengal were initiated during 1952-57 at Andhra University under the leadership of Dr. E.C. La Fond, a U.S. Visiting Professor in Oceanography. The studies including physical, chemical, biological and geological aspects were published in the form of Memoirs in Oceanography (Anonymous, 1954; 1958). Measurements of photosynthetic productivity in the Bay of Bengal were made for the first time during Galathea Expedition (Steemann Nielsen and Aabery Jensen, 1957).

The next major effort in the development of Oceanography in India was the International Indian Ocean Expedition (IIOE). A comprehensive programme including fish and plankton distribution, primary production and benthic populations, in addition to basic physical and chemical parameters of the Arabian Sea and the Bay of Bengal was undertaken on INS KISTNA (1963-65). The voluminous data, collected in IIOE was sorted out and published in several volumes of collected reprints by UNESCO (IIOE, 1965-72), an Atlas on Phytoplankton production (Krey and Babenerd, 1976), and several volumes of Zooplankton Atlases (IOBC 1969-73). Thus the period 1950-65 firmly established Oceanography in India with a core of oceanographers trained during the IIOE. It has been further strengthened with the establishment of National Institute of Oceanography (NIO) at Goa, acquisition of modern research ships and offering marine science teaching and training programmes at several universities. Multi-disciplinary studies covering physical, chemical, biological, geological and geophysical aspects during the later half of twentieth century have helped considerably in understanding the Oceanographic features in the Indian Ocean region. Another milestone in ocean research is the establishment of Department of Ocean Development by the Government of India in 1981. This has resulted in the launching of major programmes such as Antarctica Research and intensive survey of manganese nodules in the Central Indian Ocean for future mining.

1.2.0 Marine Chemistry: An Emerging Discipline

During the early development of Oceanography, attention was focused primarily on the biological and physical oceanography, in so far as they were related to the existing fishery, with chemical oceanography playing a subordinate role of providing analytical data primarily on salinity, dissolved oxygen and nutrients. At that time, the chemistry of oceans or of the marine environment, in general, was considered a part of Physical Oceanography and of less importance than fisheries oriented research. The paucity of past research in chemistry of oceans all over the world may be largely attributed to the inadequacy of even simple and reliable analytical methods available at that time for chemical analysis. This was alleviated by the development of inductive salinometer for the determination of salinity and *in situ* devices (sensors) for the continuous profiling of temperature, dissolved oxygen, nutrients, etc. Rapid advances in technology during and following World War II (1940–1945) led to the development of sophisticated instrumental methods of chemical analysis. Consequent developments and refinements of the methods resulted in coping with the problems unique to ocean chemistry– namely the sensitivity required to determine extremely low concentration (ppb and below) levels of trace inorganic and organic constituents of sea water and the high precision required to measure minute variation in concentrations of major and minor components.

Marine chemistry or probably chemical oceanography is the study of chemical properties and interactions of substances present in the marine environment. The ocean is a dynamic chemical system which may be visualized as some sort of a reaction vessel containing a slightly alkaline, moderately concentrated aqueous solution of both inorganic and organic constituents in intimate contact with reactive transition metal ions, dissolved oxygen, microorganisms and catalytically active solid phases with the surface exposed to actinic sunlight (Sillen, 1961). Further, chemical reactions in ocean are largely determined by phenomena which occur at the air-sea and the water-sediment interfaces. There are several diverse areas of investigations by which marine chemistry or chemical oceanography can claim its right as an emerging discipline. These include:

1. Chemical composition and its variation in sea water, biomass and sediments.
2. Thermodynamic and kinetic aspects of several chemical equilibria in the sea.
3. Biological utilisation of chemicals as well as their regeneration (cycling) in the water column and in sediments,
4. Chemical aspects of pollution and its prevention/control in marine environment; treatment and disposal of wastes.
5. Extraction and economic recovery of chemicals, pharmaceuticals, minerals, oil and gas, and energy from the sea.
6. Applications of radioisotopes in oceanography, and
7. Development of reliable new analytical techniques and methods.

1.3.0 Gross Chemical Composition of Sea water

Sea water is a complex mixture consisting, on average 96.5 per cent of water, 3.5 per cent of salts and a few parts per million (ppm) of organic constituents and particulate matter. About 90 naturally occurring elements have so far been identified and evaluated in sea water. The remainder are likely to be detected when more sophisticated and sensitive analytical methods are available. The elements so far determined in the dissolved form show a wide range of concentrations as shown in Table 1.1.

The dissolved components in sea water can be sub-divided into (*i*) major and minor elements (*ii*) dissolved gases (*iii*) nutrients and (*iv*) radioactive nuclides. Further, the dissolved fraction also contains (not shown in Table 1.1) several organic constituents. Similar to dissolved components, a wide variety of suspended solids also exist in sea water comprising both inorganic and organic constituents such as clays, minerals, living and dead phyto and zooplankton, bacteria and other organisms. The dissolved and particulate fractions can conveniently be distinguished and separated from each other by filtration through a membrane or glass fibre filter having a pore size(diameter) of 0.45 micrometres (0.45×10^{-6} m). The fraction retained on the filter is designated as particulate and that passed through the membrane as dissolved fraction.

Table 1.1: Average Abundance of Chemical Elements in Sea water

Element		*Concentration (mg.l^{-1} i.e. Parts per million, ppm)*	*Some Probable Dissolved Chemical Species*	*Total Amount in the Oceans (tonnes)*
Chlorine	Cl	1.95×10^{4}	Cl^-	2.57×10^{16}
Sodium	Na	1.077×10^{4}	Na^+	1.42×10^{16}
Magnesium	Mg	1.290×10^{3}	Mg^{2+},$MgSO_4$,$MgCO_3$	1.71×10^{15}
Sulphur	S	9.05×10^{2}	SO_4^{2-}, $NaSO_4^+$	1.2×10^{15}
Calcium	Ca	4.12×10^{2}	Ca^{2+}	5.45×10^{14}
Potassium	K	3.80×10^{2}	K^+	5.02×10^{14}
Bromine	Br	67	Br^-	$8.86 \mathrm{X} 10^{13}$
Carbon	C	28	HCO_3^-	3.7×10^{13}
Nitrogen	N	11.5	N_2 gas, NO_3^-, NH_4^+	1.5×10^{13}
Strontium	Sr	8	Sr^{2+}	1.06×10^{13}
Oxygen	O	6	O_2	7.93×10^{12}
Boron	B	4.4	$B(OH)_3$,$B(OH)_4^-$, $H_2BO_3^-$	5.82×10^{12}
Silicon	Si	2	$Si(OH)_4$	2.64×10^{12}
Fluorine	F	1.3	F^-, MgF^+	1.72×10^{12}
Argon	Ar	0.43	Ar gas	5.68×10^{11}
Lithium	Li	0.18	Li^+	2.38×10^{11}
Rubidium	Rb	0.12	Rb^+	1.59×10^{11}
Phosphorus	P	6×10^{-2}	HPO_4^{2-},PO_4^{3-},$H_2PO_4^-$	7.93×10^{10}
Iodine	I	6×10^{-2}	IO_3^-, I^-	7.93×10^{10}
Barium	Ba	2×10^{-2}	Ba^{2+}	2.64×10^{10}
Molybdenum	Mo	1×10^{-2}	MoO_4^{2-}	1.32×10^{10}
Uranium	U	3.2×10^{-3}	$UO_2(CO_3)_2^{4-}$	4.23×10^{9}
Vanadium	V	2×10^{-3}	$H_2VO_4^-$, HVO_4^{2-}	2.64×10^{9}
Arsenic	As	2×10^{-3}	$HAsO_4^{2-}$,$H_2AsO_4^-$	2.64×10^{9}

Contd...

Table 1.1–Contd...

Element		Concentration (mg.l^{-1} i.e. Parts per million, ppm)	Some Probable Dissolved Chemical Species	Total Amount in the Oceans (tonnes)
Titanium	Ti	$1x10^{-3}$	$Ti(OH)_4$	$1.32x10^{9}$
Zinc	Zn	$5x10^{-4}$	$ZnOH^+$, Zn^{2+},$ZnCO_3$	$6.61x10^{8}$
Nickel	Ni	$4.8x10^{-4}$	Ni^{2+}, $NiCO_3$, $NiCl^+$	$6.35x10^{8}$
Aluminium	Al	$4x10^{-4}$	$Al(OH)_4^-$	$5.29x10^{8}$
Caesium	Cs	$4x10^{-4}$	Cs^+	$5.29x10^{8}$
Chromium	Cr	$3x10^{-4}$	$Cr(OH)_3$,CrO_4^{2-}, $NaCrO_4^-$	$3.97x10^{8}$
Antimony	Sb	$2x10^{-4}$	$Sb(OH)_6^-$	$2.64x10^{8}$
Krypton	Kr	$2x10^{-4}$	Kr gas	$2.64x10^{8}$
Selenium	Se	$2x10^{-4}$	SeO_3^{2-}, SeO_4^{2-}	$2.64x10^{8}$
Neon	Ne	$1.2x10^{-4}$	Ne gas	$1.59x10^{8}$
Cadmium	Cd	$1x10^{-4}$	$CdCl_2$	$1.32x10^{8}$
Copper	Cu	$1x10^{-4}$	$CuCO_3$,$Cu(OH)^+$,Cu^{2+}	$1.32x10^{8}$
Tungsten	W	$1x10^{-4}$	WO_4^{2-}	$1.32x10^{8}$
Iron	Fe	$5.5x10^{-5}$	$Fe(OH)_2^+$,$Fe(OH)_4^-$	$7.27x10^{7}$
Xenon	Xe	$5x10^{-5}$	Xe gas	$6.61x10^{7}$
Manganese	Mn	$3x10^{-5}$	Mn^{2+},$MnCl^+$	$3.97x10^{7}$
Zirconium	Zr	$3x10^{-5}$	$Zr(OH)_4$	$3.97x10^{7}$
Niobium	Nb	$1x10^{-5}$	$Nb(OH)_6^-$	$1.32x10^{7}$
Thallium	Tl	$1x10^{-5}$	Tl^+	$1.32x10^{7}$
Thorium	Th	$1x10^{-5}$	$Th(OH)_4$	$1.32x10^{7}$
Hafnium	Hf	$7x10^{-6}$	$Hf(OH)_5^-$	$9.25x10^{6}$
Helium	He	$6.8x10^{-6}$	He gas	$8.99x10^{6}$
Germanium	Ge	$5x10^{-6}$	$Ge(OH)_4$,$H_3GeO_4^-$	$6.61x10^{6}$
Rhenium	Re	$4x10^{-6}$	ReO_4^-	$5.29x10^{6}$
Cobalt	Co	$3x10^{-6}$	Co^{2+}	$3.97x10^{6}$
Lanthanum	La	$3x10^{-6}$	$La(OH)_3$,La^{3+},$LaCO_3^+$	$3.97x10^{6}$
Neodymium	Nd	$3x10^{-6}$	$Nd(OH)_3$,$NdCO_3^+$,Nd^{3+}	$3.97x10^{6}$
Cerium	Ce	$2x10^{-6}$	$Ce(OH)_3$,$CeCO_3^+$,Ce^{3+}	$2.64x10^{6}$
Lead	Pb	$2x10^{-6}$	$PbCO_3$, $Pb(CO_3)_2^{2-}$, Pb^{2+}	$2.64x10^{6}$

Contd...

Table 1.1–Contd...

Element		Concentration ($mg.l^{-1}$ i.e. Parts per million, ppm)	Some Probable Dissolved Chemical Species	Total Amount in the Oceans (tonnes)
Silver	Ag	$2x10^{-6}$	$AgCl_2^-$	$2.64x10^6$
Gallium	Ga	$2x10^{-6}$	$Ga(OH)_4^-$	$2.64x10^6$
Tantalum	Ta	$2x10^{-6}$	$Ta(OH)_5$	$2.64x10^6$
Yttrium	Y	$1x10^{-6}$	YCO_3^+, Y^{3+}	$1.32x10^6$
Mercury	Hg	$1x10^{-6}$	$HgCl_4^{2-}, HgCl_2$	$1.32x10^6$
Dysprosium	Dy	$9x10^{-7}$	$Dy(OH)_3, DyCO_3^+, Dy^{3+}$	$1.19x10^6$
Erbium	Er	$8x10^{-7}$	$Er(OH)_3, ErCO_3^+, Er^{3+}$	$1.06x10^6$
Ytterbium	Yb	$8x10^{-7}$	$Yb(OH)_3, YbCO_3^+$	$1.06x10^6$
Gadolinium	Gd	$7x10^{-7}$	$Gd(OH)_3, GdCO_3^+, Gd^{3+}$	$9.25x10^5$
Praseodymium	Pr	$6x10^{-7}$	$Pr(OH)_3, PrCO_3^+, Pr^{3+}$	$7.93x10^5$
Samarium	Sm	$6x10^{-7}$	$Sm(OH)_3, SmCO_3^+, Sm^{3+}$	$7.93x10^5$
Tin	Sn	$6x10^{-7}$	$SnO(OH)_3^-$	$7.93x10^5$
Scandium	Sc	$6x10^{-7}$	$Sc(OH)_3$	$7.93x10^5$
Holmium	Ho	$3x10^{-7}$	$Ho(OH)_3, HoCO_3^+, Ho^{3+}$	$3.97x10^5$
Beryllium	Be	$2x10^{-7}$	$BeOH^+$	$2.64x10^5$
Lutetium	Lu	$2x10^{-7}$	$Lu(OH)^{2+}, LuCO_3^+$	$2.64x10^5$
Europium	Eu	$2x10^{-7}$	$Eu(OH)_3, EuCO_3^+, Eu^{3+}$	$2.64x10^5$
Indium	In	$2x10^{-7}$	$In(OH)_2^{2+}, In(OH)_3^+$	$2.64x10^5$
Thulium	Tm	$2x10^{-7}$	$Tm(OH)_3, TmCO_3^+, Tm^{3+}$	$2.64x10^5$
Terbium	Tb	$1x10^{-7}$	$Tb(OH)_3, TbCO_3^+, Tb^{3+}$	$1.32x10^5$
Palladium	Pd	$5x10^{-8}$	$Pd^{2+}, PdCl^+$	$6.61x10^4$
Gold	Au	$2x10^{-8}$	$AuCl_2^-$	$2.64x10^4$
Bismuth	Bi	$2x10^{-8}$	$Bi0^+, Bi(OH)_2^+$	$2.64x10^4$
Tellurium	Te	$1x10^{-8}$	$Te(OH)_6$	$1.32x10^4$
Radium	Ra	$7x10^{-11}$	Ra^{2+}	92.5
Protactinium	Pa	$5x10^{-11}$	—	66.1
Radon	Rn	$6x10^{-16}$	Rn gas	$7.93x10^{-4}$
Polonium	Po	$5x10^{-16}$	$Po^{2+}, Po(OH)_2$	$6.61x10^{-4}$

Source: Open University, 1995.

1.4.0 Sources, Cycles and Sinks of Dissolved and Particulate Matter in the Sea

It is generally believed that rivers are the principal source of dissolved salts in sea water. If it is true, the sea water is simply a more concentrated form of river water. Further, HCO_3^- rather than Cl^- would be the principal anion, and Ca^{2+} rather than Na^+, the principal cation in sea water. However, rivers pick up salts during the weathering on land and precipitate some of them in different amounts, and at different rates by the time they drain into the sea. A comparison of the average concentration of principal dissolved constituents in rain water, river water and sea water as shown in Figure 1.1, suggests unambiguously that sea water is not simply a more concentrated form of river water. It is evident from Figure 1.1 that

1. Total dissolved solids (TDS) follow the order,

   Sea water >> river water > rain water.

2. Composition wise sea water is more akin to rain water than river water. This is evident from relatively high content of Na^+, K^+ and Cl^- ions in sea water and rain water than in river water. On the other hand, the content of Ca^{2+}, and HCO_3^- ions in river water is higher than in sea water and rain water.

This is explained on the basis of addition of these constituents to rain water during chemical weathering of rocks on land. Rain water contains dissolved gases, particularly, CO_2 and SO_2 and hence it is slightly acidic (pH 5.7). When rain falls on land, the acidity is neutralized by reaction with minerals in soils and rocks which can be represented by the following typical examples:

$$CaCO_3(s) + CO_2 + H_2O \longrightarrow Ca^{2+} + 2HCO_3^- \qquad 1.1$$

$$\underset{\text{(Albite)}}{2NaAlSi_3O_8} + 2CO_2 + 3H_2O \longrightarrow \underset{\text{(Kaolinite)}}{Al_2Si_2O_5(OH)_4} + 2Na^+ + 2HCO_3^- + 4SiO_2 \qquad 1.2$$

Exceptionally high concentrations of Ca^{2+} and HCO_3^- in river water are due to the production of these ions during weathering of carbonate and alumino silicate minerals commonly occurring in sedimentary and igneous/metamorphic rocks by the above reactions. However, chloride content in river water cannot be explained on the

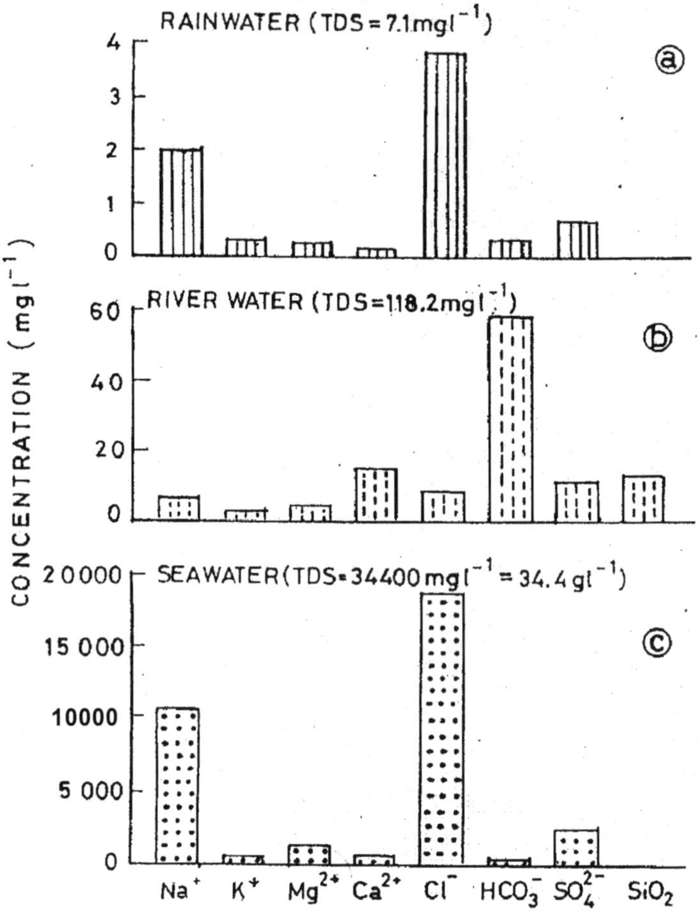

Figure 1.1: The Average Composition of Rain, River and Sea Water (*Source*: Open University, 1995)

basis of chemical weathering, because the average chloride of continental crustal rocks is of the order of 0.01 per cent and so only very minute proportion of it in river water comes from weathering. It follows that virtually all chloride in river water must have been contributed by sea salts (NaCl) recycled via oceanic aerosols. Hence it can be stated that some of the salts in river water come from chemical weathering of source rocks and the remainder being recycled sea salts contributed by rain.

It is a simple matter that the origin of Na^+, K^+, Ca^{2+} and Mg^{2+} in sea water can be attributed to rock weathering since these elements have high abundance in the earth' s crust. By contrast and as stated earlier, only a negligible proportion of Cl^-, comes from weathering. Then what is its origin in sea water? The answer lies in volcanism. HCl is an important constituent of volcanic gases. As earth in its early history was very hot, large quantities of HCl gas emitted and quickly dissolved in oceans. Thus, chloride is classified as an excess volatile, a constituent of sea water that cannot be accounted for by rock weathering. Similarly SO_2, HBr and volatile boron compounds are all known to be components of volcanic gases, along with CO_2, nitrogen, argon, hydrogen and water vapour. In other words, one can suppose that the world' s water inventory is the result of planetary degassing. Further, hydrothermal activity at ocean ridges and other sites of oceanic volcanism is known to supply some elements like Ca^{2+}, Mg^{2+} to sea water.

We have discussed so far the major sources of dissolved constituents. The major sources of particulate matter in oceans are (i) rivers carrying particulates in suspension (silt and clay), (ii) wind borne dust (quartz, diatom skeletons, organic detritus and volcanic ashes), and (iii) biogenic particulates resulting from primary and secondary production (skeletal remains, feacal pellets, detritus, etc.,).

1.5.0 Geochemical Balance and Residence Times of Elements in Sea Water

The present ionic composition of sea water results from a balance between the rate at which the dissolved matter is added to the ocean from land and atmosphere, and the rate at which it is removed from the sea by incorporation into the sediments or by being returned to the atmosphere. It is the geochemical balance which appears to have kept the relative composition of sea water fairly constant. This implies that the oceans are chemically in a steady state as a result of the rate of addition of dissolved constituents to sea water being balanced by the rate of removal, so that their concentrations do not change significantly with time. This is evident from the fact that the composition of sea water has not varied significantly over the past several hundred million years.

Based on the steady state, a useful concept known as the residence time ( ) has been developed to characterize dissolved constituents in sea water. This is defined as the average time a dissolved constituent remains in the sea water before being removed completely by the processes mentioned above. The residence time can be expressed as:

$$= \frac{\text{Total mass of the dissolved constituent in oceans}}{\text{Its rate of supply or removal}} \qquad 1.3$$

Since the denominator is usually expressed as annual rate, residence times are normally expressed in years.

For residence time calculations, it is assumed that rivers are the only source of supply of dissolved constituents. This is still valid enough for most dissolved constituents to provide reasonable estimates of residence times, even though it is now known that hydrothermal solutions provide significant sources of some elements. The residence time of an element in sea water can, therefore, be estimated by dividing its mass in the oceans by its annual input from rivers. The annual flux of each element into the oceans via rivers can readily be calculated from the total annual inflow of river water to the oceans multiplied by average concentration of that element in river water. The total mass of each element in the ocean is also easily calculated from data on the average concentration in sea water and the total mass of water in the oceans. Oceanic residence times of some elements are listed in Table 1.2. It should be noted that the residence times are only approximate. The averages used conceal wide variations and in many cases are based on limited data apart from the basic assumptions which are not wholly valid as mentioned earlier.

Residence time of an element in the sea apparently is related to its chemical behaviour. Elements like Na, K, Cl, etc., which are little affected by sedimentary or biological processes have long residence times, remain in sea water of the order of 10^6–10^8 years. On the other hand, minor or trace elements have short residence times and are removed from the sea water in 10^2–10^4 years or even less. Dissolved constituents with less residence times are often said to be more reactive than others having long residence times in sea water.

Table 1.2: River Fluxes and Residence Times of Some Dissolved Constituents in Sea water

Constituent	*River Flux ($x10^8$ t $year^{-1}$)*	*Mass in Ocean ($x10^{14}$ t)*	*Residence Time ($x10^6$ year)*
Na^+	2.05	144	70.2
K^+	0.75	5	6.7
Ca^{2+}	4.88	6	1.23
Mg^{2+}	1.33	19	14.3
Cl^-	2.54	261	103
HCO_3^-	18.95	1.9	0.1
SO_4^{2-}	3.64	37	10.2
SiO_2	4.26	0.08	0.02
Fe	0.22	0.000014	0.00006
Mn	0.001	0.00002	0.0002
Cu	0.0007	0.000021	0.03
Co	0.001	0.000001	0.0001
Zn	0.0007	0.000042	0.006

Source: Open University,1995.

We can calculate also residence time of water in the oceans. There is a net removal of water (336×10^{15} kg) due to evaporation from the ocean surface in each year. This water falls on the land and return to the ocean as river runoff, which is known as hydrological cycle as shown in Figure 1.2 and Table 1.3. Precipitation in the ocean adds 300×10^{15} kgs and run off from land adds 36×10^{15} kgs per year so that the total ($336 x 10^{15}$ kgs) balances the amount lost by evaporation. The oceans contain $13{,}22{,}000 \times 10^{15}$ kgs of water. Hence the residence time of water in the oceans, calculated from equation 1.3 is:

$$\text{Residence time of water} = \frac{13{,}22{,}000 \times 10^{15}}{336 \times 10^{15}} \sim 3{,}900 \text{ years}$$

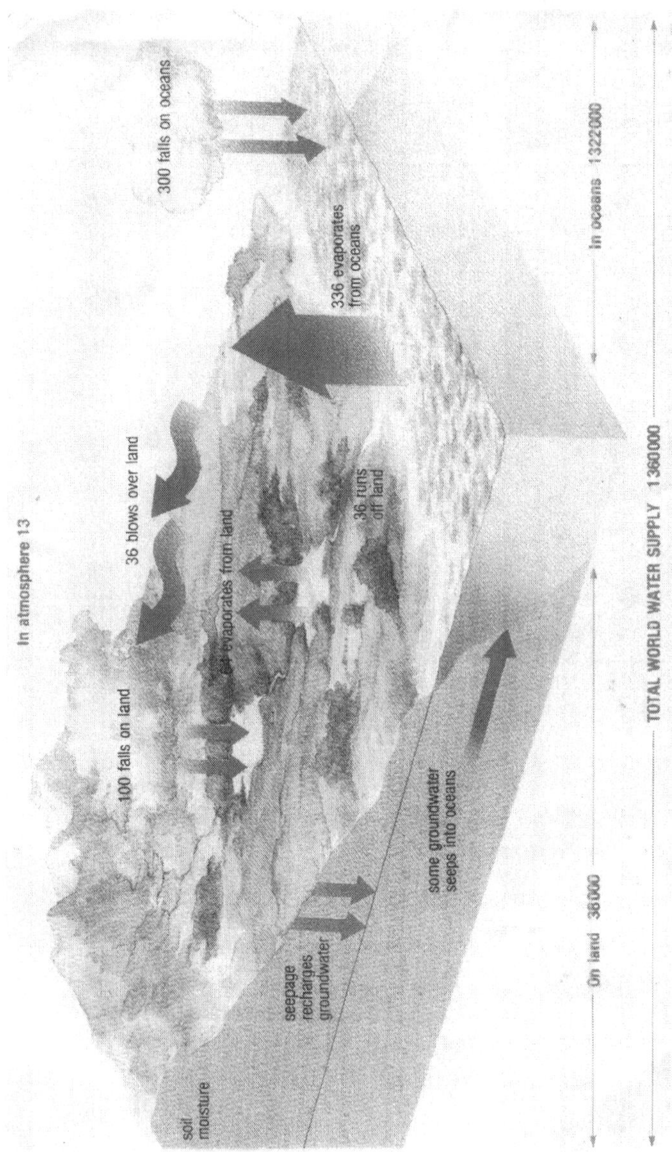

Figure 1.2: The Hydrological Cycle. All quantities shown are 10^{15} kg = 10^3 km^3 Water (*Source*: Open University, 1995)

Table 1.3: Waters on Land (x 10^{15} kg)

Rivers and streams	1
Freshwater lakes	125
Salt lakes and inland seas	104
Total surface water	230
Glaciers and ice-caps	29,300
Soil moisture and seepage	70
Groundwater	8,400
Total on land	38,000

Source: Open University,1995.

1.6.0 The Concept of Salinity

Since oceans are in a steady state the dissolved salts in sea water have a nearly constant composition. We can therefore use the most abundant constituent namely, chloride as the index of the amount of dissolved salts present in a given amount of sea water. Thus, chlorinity is defined as the amount of chlorine in grammes present in one kilogramme of sea water (while bromine and iodine are replaced by chlorine). It is expressed as parts per thousand or parts per milli (‰). Since chlorinity indicates only the amount of halides ($Cl^- + Br^- + I^-$), another term salinity was introduced to represent the total amount of dissolved salts in grammes present in one kg of sea water (iodine and bromine are replaced by chlorine and all organic matter is oxidized). Salinity is also expressed as parts per thousand (‰). The salinity (S‰) is related to chlorinity (Cl‰) by equation.

$$\text{Salinity (S‰)} = 1.80655 \times \text{Chlorinity (Cl‰)} \qquad 1.4$$

Chlorinity was measured by chemical methods (argentometric titration) and salinity was computed using equation 1.4. This is the only method widely used for the evaluation of salinity until the mid 1960s. However, this method is superceded by conductivity measurements since salinity of sea water was found to be proportional to its electrical conductivity. A new empirical relationship was proposed in 1966 for salinity based on the equation.

$$\text{Salinity } (‰) = -0.08996 + 28.29720\, R_{15} + 12.80832\, (R_{15})^2 - 10.67869\, (R_{15})^3 + 5.98624\, (R_{15})^4 - 1.32311 (R_{15})^5 \qquad 1.5$$

where,

R_{15} is the conductivity ratio which is defined as:

$$R_{15} = \frac{\text{Conductivity of sea water sample}}{\text{Conductivity of standard sea water } (S=35.000‰)} \qquad 1.6$$

at 15^0 C and under a pressure of standard (one) atmosphere.

Tables are used for direct conversion of R_{15} to S‰ and for converting conductivity ratio at temperature and pressure of measurement other than 15°C and 1 atm. pressure into R_{15}.

The above equation has been modified in 1980 and the practical salinity of sea water is related to conductivity ratio (K_{15}) which is defined as:

$$K_{15} = \frac{\text{Conductivity of sea water sample}}{\text{Conductivity of standard KCl solution}} \qquad 1.7$$

at 15°C and 1 atm pressure, the concentration of the standard KCl solution being 32.4356 g kg^{-1}.

The practical salinity is related to the ratio, K_{15} by the equation.

$$S = 0.0080 - 0.1692\, (K_{15})^{1/2} + 25.3851\, (K_{15}) + 14.0941\, (K_{15})^{3/2} - 7.0261\, (K_{15})^2 + 2.7081\, (K_{15})^{5/2} \qquad 1.8$$

If by definition $K_{15} = 1$, the above equation simplifies to

$$S = 0.0080 - 0.1692 + 25.3851 + 14.0941 - 7.0261 + 2.7081 \qquad 1.9$$

and the practical salinity exactly equals to 35.000.

Since the definition of salinity is related to conductance ratio, salinities are now a days presented as simple numbers. It is important, however, to note that the number (psu) represents grammes per kilogramme or parts per thousand ($x10^{-3}$). As in the case of R_{15}, tables or computer algorithms are used for the direct conversion of K_{15} into S and conductivity ratios measured at temperatures and pressures other than 15°C and 1 atm pressure into K_{15}.

Salinity based on the conductivity measurement depends on temperature and pressure at which it is determined. Hence it slightly deviates from the definition of salinity being the total dissolved salt content of sea water. In fact for open ocean sea water the two are closely related, the total dissolved salt content in one kg of sea water is equal to 1.00510 × S, where S is defined by equation 1.8.

1.6.1 Determination of Salinity

One of the obvious ways of measuring the salinity is to take a known amount of sea water, evaporate it to dryness and then weigh the remaining salts gravimetrically. Although simple, this method gives variable and unreliable results because of several reasons. The residue left after evaporation is a complex mixture of salts, together with water of hydration bound to the salts and a small amount of organic material. Even though the amount of water left behind can be reduced to a large extent by heating the residue at elevated temperature (120°C to 150°C), this leads to other problems such as (*i*) decomposition of small amounts of salts (hydrous $MgCl_2$ crystals with a loss of HCl) and, (*ii*) Vapourisation and decomposition of the organic matter, thus leading to variable results in salinity depending upon the conditions employed to drive off the water. Apart from this, gravimetric determination of salinity is both difficult and tedious and so other methods have been investigated.

One of the methods proposed for the determination of salinity is the measurement of concentration of halides (chloride + bromide + iodide), in other words chlorinity of sea water. Chlorinity is determined by titrating a known amount of sea water, with silver nitrate solution of known strength using potassium chromate(K_2CrO_4) as indicator. As the Ag+ ions are added they react with the halide ions to form insoluble precipitates of silver halides.

$$Ag^+ + Cl^- \longrightarrow AgCl(s) \qquad 1.10$$

$$Ag^+ + Br^- \longrightarrow AgBr(s) \qquad 1.11$$

$$Ag^+ + I^- \longrightarrow AgI(s) \qquad 1.12$$

when all the halide ($Cl^- + Br^- + I^-$) ions are precipitated from an aliquot of sea water, the addition of excess Ag^+ ions will change the colour of yellow potassium chromate (K_2CrO_4) to brick red silver chromate (Ag_2CrO_4) thus signaling the equivalence (end) point in the titration.

$$K_2CrO_4 + 2Ag^+ \longrightarrow Ag_2CrO_4(s) + 2K^+ \quad 1.13$$
(Yellow) (Brick red)

The volume of silver nitrate required to react with halides is directly proportional to the chlorinity of the sample. Substitution of measured chlorinity in equation 1.4 yields salinity of the sample. Standard sea water (chlorinity = 19.375‰or salinity = 35.000‰) is used for standardizing silver nitrate solution.

Though pure water is a bad conductor of electricity, sea water is a good conductor because of the presence of dissolved salts (electrolytes) like NaCl, KCl, $MgCl_2$, etc. In 1930' s it was established that the electrical conductivity is proportional to its salinity. The resistance of sea water which actually to be determined experimentally, is inversely proportional to the conductivity. Based on this principle, conductivity bridges or salinometers constructed using the simple electrical circuits as in wheatstone' s Bridge (Figure 1.3) are employed for the determination of chlorinity of sea water.

In operation, the variable resistance (R_3), is adjusted keeping (R_4) fixed until the current flowing through the galvanometer falls to zero, when the following relationship holds:

$$\frac{R_1}{R_2} = \frac{R_3}{R_4} \quad 1.14$$

As values of all the resistances, except R_2 are known, the resistance of the unknown sea water sample (R_2) can be calculated. A standard sea water of salinity 35.000 psu is used as R_1 for calibration purpose. Commercial instruments known as inductive salinometers are now available to accurately measure salinity of sea water.

Conductivity depends to a large extent on temperature of the solution and to a small extent on pressure. Physical oceanography studies such as characterization of water masses, determination of vertical stability and calculation of geostrophic current flow ideally require salinity measurement to an accuracy of ± 0.0001 if possible. To achieve this, conductivity must be measured to 1 part in 40,000. A change of this magnitude can be induced by a temperature change of only 0.001°C. So careful control of temperature of sea water

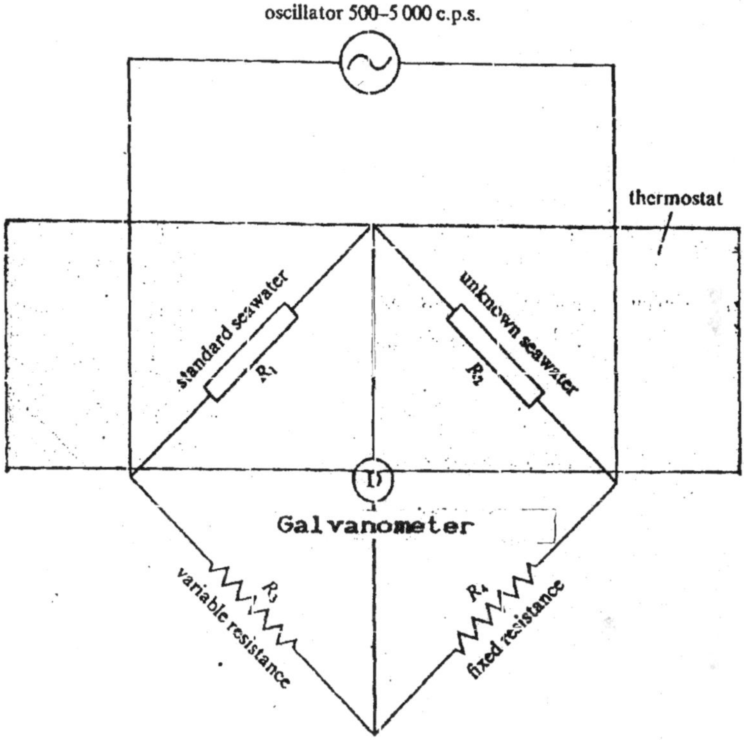

Figure 1.3: A Simple Bridge Circuit for Measuring Conductivity of Seawater (*Source*: Open University, 1989)

sample and the standard should be required. Modern salinometers can measure salinity to ± 0.003 psu or better. Conductivity sensors have been incorporated into insitu temperature-salinity instruments for use in shallow waters, and into conductivity-temperature-depth (CTD) probes for use in the deep oceans.

1.6.2 Spatial Distribution of Salinity

The salinity of surface waters of oceans is at its maximum at latitudes of about 20°, where evaporation exceeds precipitation. These regions correspond to the hot barren deserts that exist on similar latitudes on land. Salinities decrease both towards higher latitudes and towards equator (Figure 1.4a). Local modifications are superimposed on this regional pattern, particularly near landmasses. Surface salinity may be reduced by an influx of fresh water at the mouths of large rivers like Ganges, Brahmaputra, Indus, Amazon etc., and by melting ice and snow at high latitudes. On the other hand, surface salinities tend to be high in lagoons and other partly enclosed shallow marine basins (Red Sea and Persian Gulf) at low latitudes, where evaporation is high and the inflow of water from the adjacent land areas is limited. Surface salinities are controlled by the balance between evaporation and precipitation. Figure 1.4(b) shows their close relationship– low salinity where precipitation exceeds evaporation and vice versa. The salinity minimum along the equatorial belt is due to an excess rainfall over evaporation, despite high average insolation and temperatures.

1.6.3 Salinity Distribution in the Indian Ocean

The Indian Ocean north of the equator chiefly comprises the Arabian Sea and the Bay of Bengal. The Arabian Sea is bordered on the northern, eastern and western sides by the landmasses of Asia and Africa. It is connected to the Persian Gulf through the Gulf of Oman by the shallow Hormuz Strait. No sill or saddle separates the Persian Gulf from the Gulf of Oman. Similarly a 125 m deep sill at the Strait of Bab-el-Mandab separates the Red Sea from the Arabian Sea through the Gulf of Aden. Total area of the Arabian Sea between latitudes 0° and 25° N and longitudes 50° and 80° E is about 6.225×10^6 km^2 (Qasim, 1977). It is an area of negative water balance where evaporation exceeds precipitation and runoff. This results in the formation of several high salinity areas as shown in Figure 1.5. In contrast to the Arabian Sea, the Bay of Bengal which occupies an area of 4.087×10^6 km^2 between 0° and 23° N and 80° and 100° E (Qasim. loc. cit) is a region of positive water balance. The total annual river run off to the Bay of Bengal has been estimated to be around 2700 km^3 (Sen Gupta, 2001). The net positive water balance and the massive river run off result in low surface salinities especially in the northern Bay of Bengal. Further, surface salinity is lower at any position in the Bay of Bengal than the Arabian Sea (Figure 1.5). Waters of highest salinity appear in the northern Arabian Sea with the lowest salinity in the western, northern and eastern parts of the Bay of Bengal. Low salinity waters from the southeast Indian Ocean

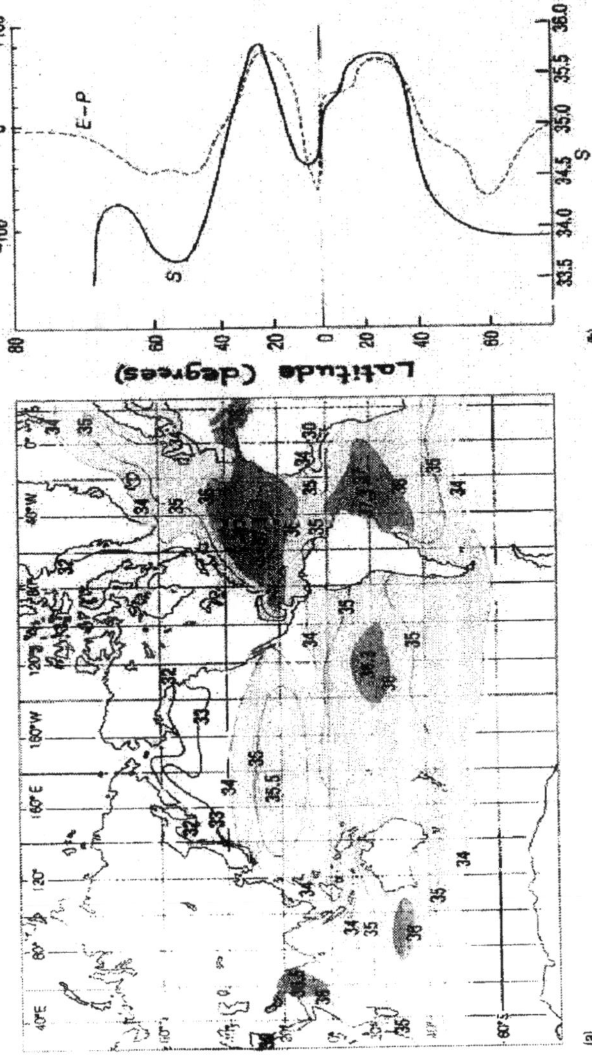

Figure 1.4: (a) The Approximate Positions of Mean Annual Isohalines (b) Average Values of Salinity (Solid Line) and the Difference Between Average Annual Evaporation and Precipitation (E–P) (Dotted Line), Plotted Against Latitude (*Source:* Open University, 1995)

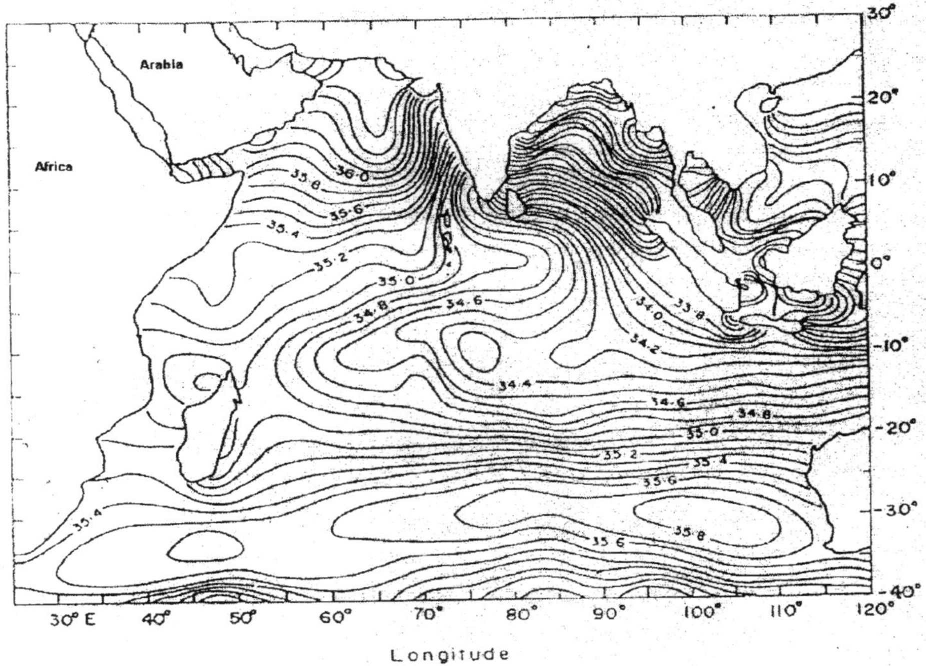

Figure 1.5: Annual Mean Salinity Distribution at Surface of the Indian Ocean Contour Interval 0.1 Psu (Source: Murty and Murthy, 2001)

flows with southwest coast of India. The salinity (isohalines) gradually increases from 35.0 to 35.6 psu south of the equator (25°–35°).

Multiple Choice Questions

1. Which of the following relative cation abundance pattern apply to sea water?
   (a) $Na^+ > Mg^{2+} > Ca^{2+}$ (b) $Ca^{2+} > Mg^{2+} > K^+$
   (c) $K^+ > Ca^{2+} > Na^+$ (d) $Mg^{2+} > Na^+ > K^+$ ( )

2. Which of the following has closer similarity to rain water?
   (a) Sea water (b) River water
   (c) Ground water (d) None of them ( )

3. Approximately how many times rain water is dilute than sea water with respect to TDS?
   (a) 5000 (b) 300
   (c) 200 (d) 100 ( )

4. Approximately how many times river water is dilute than sea water with respect to TDS?
   (a) 500 (b) 300
   (c) 100 (d) 400 ( )

5. Which of the following relative anion abundance pattern apply to sea water?
   (a) $HCO_3^- > SO_4^{2-} > Cl^-$ (b) $Cl^- > SO_4^{2-} > HCO_3^-$
   (c) $SO_4^{2-} > Cl^- > HCO_3^-$ (d) None of the above ( )

6. Relatively high concentration of Ca^{2+} and HCO_3^- in river water is due to
   (a) Precipitation from atmosphere
   (b) Weathering of rocks on land
   (c) Recycling of sea salt via atmosphere
   (d) None of the above ( )

7. The principal origin of chloride in sea water is
   (a) River water
   (b) Recycling of sea salts via rivers

(c) Volcanic activity

(d) Hydrothermal Activity ( )

8. Approximate residence time of water (in years) in ocean is

(a) 4×10^3 (b) 4×10^6

(c) 1×10^2 (d) 4×10^4 ( )

9. Which of the following relative cation abundance pattern apply to river water?

(a) $Na^+ > Mg^{2+} > Ca^{2+}$ (b) $Ca^{2+} > Na^{2+} > Mg^{2+}$

(c) $Na^+ > Ca^{2+} > Mg^{2+}$ (d) $K^+ > Mg^{2+} > Na^+$ ( )

10. Which of the following relative anion abundance pattern applies to river water?

(a) $HCO_3^- > SO_4^{2-} > Cl^-$ (b) $Cl^- > SO_4^{2-} > HCO_3^-$

(c) $SiO_4^{4-} > SO_4^{2-} > HCO_3^-$ (d) None of the above ( )

11. Which of the following statement is false

(a) The approximate percentage of salt in sea water is 3.5

(b) Dissolved constituents in sea water are separated from particulate constituents by filtration through 0.45 filter.

(c) Rivers are the principal source of dissolved salts of sea water.

(d) Sea water composition resembles more of river water than rain water.

( )

12. Which of the following statements is correct with regard to the distribution of surface salinity in the oceans?

(a) Maximum at equator

(b) Increases towards higher latitudes.

(c) Maximum at about 20° latitude.

(d) Maximum in isolated regions (Lagoons) where the rate of precipitation is greater than evaporation.

( )

13. In practical salinity definition, the ratio of conductance (K_{15}) is defined as

    (a) $\frac{\text{Conductance of sample sea water}}{\text{Chlorinity of standard sea water}}$

    (b) $\frac{\text{Conductance of sample sea water}}{\text{Salinity of standard sea water}}$

    (c) $\frac{\text{Conductance of sample sea water}}{\text{Conductance of standard KCl solution}}$

    (d) None of the above. ( )

14. Conductance of sea water depends upon

    (a) Temperature

    (b) Pressure

    (c) Temperature and Pressure

    (d) None of the above ( )

15. Elements having relatively high residence time (10^6 to 10^8 years) are characterised in sea water by

    (a) High reactivity (b) Low reactivity

    (c) No reactivity (d) None of the above. ( )

Short Answer Questions

1. Calculate the salinity of sea water with chlorinity of 19‰, 21 ‰

2. Calculate the residence time of sodium in sea water (mass of dissolved Na in oceans = 144×10^{14} tonnes and its river flux to world oceans = 2.05×10^8 tonnes Y^{-1}).

3. What is the relation between chlorinity and salinity of sea water?

4. How is practical salinity defined?

5. In what way the present practical salinity differs from the definition of earlier salinity based on the conductance measurement?

6. What is the principle involved in estimation of salinity based on chemical methods?

7. Explain the terms positive and negative water balance of seas? Illustrate each with one typical example.

8. From the Figure 1.2, indicate
   (a) What is the rate of evaporation from the oceans?
   (b) Is it balanced from the combined amounts of precipitation and run off?
   (c) What is the amount of water that moves into the atmosphere annually?

9. How do you define chlorinity and salinity of sea water?

Review Questions

1. How the salinity of sea water determined using a salinometer? Explain the principle of the method?

2. Define residence time and discuss its significance?

3. Explain the importance of Law of the Seas.

4. Describe the important land marks in the early development of oceanography world wide.

5. What are the major milestones in the development of oceanography in India?

6. What are the important conclusions that you can arrive at by comparison of the average chemical composition of rain, river and sea waters?

7. " Marine Chemistry is an emerging discipline". How do you justify this statement.

8. Why is surface salinity maximum at latitudes around 20°?

9. Briefly describe the hydrological cycle of water.

10. Explain salient features of surface salinity distribution in the Indian Ocean.

11. How do you account for the geochemical balance of elements in sea water?

Answers to

Multiple Choice Questions

1. a	2. a	3. a	4. b	5. b	6. b	7. c	8. a
9. b	10. a	11. d	12. c	13. c	14. c	15. b	

Short Answer Questions

1. S(‰) = 1.80655 × 19 = 34.324
   = 1.80655 × 21 = 37.938

2. Residence time of sodium ($_{Na}$) =

$$\frac{144x10^{14}}{2.05 \times 10^{8}} = 7.02\ x10^{7} \text{ years}$$

3. (a) $336x10^{15}$ kg y^{-1}
   (b) yes
   (c) 400×10^{15} kg y^{-1}

Chapter 2
Major and Minor Elements

2.1.0 Major Elements

Although sea water contains species of almost all the elements in the periodic table, only fourteen (O, H, Cl, Na, Mg, S, Ca, K, Br, C, Sr, B, Si and F) have concentrations equal to or greater than, one part per million by weight. Excluding O, H and Si, the remaining II elements are defined as major constituents (as shown in Tables 2.1 and 2.2) which account for 99 per cent of the dissolved salts in the oceans. Nitrogen, oxygen and silicon are not included, despite their relatively high concentrations because the first two are dissolved gases and the third is a nutrient. By and large the elements are geochemically more un-reactive than others and hence termed as conservative. The remaining elements, which can be defined as minor ones exists in micro or submicro molar concentrations. Nevertheless, they involve in some significant biochemical reactions occurring in the marine environment, and hence termed as non-conservative elements. It can be seen from Table 2.1 that the total concentration of major elements (34.482‰) comes nearly to the average salinity (35.000 psu) of sea water. It is interesting to note that the total concentration of anions (21.861‰) is far more than the total concentration of cations (12.621‰). This is due to the fact that they are expressed by weight but not in the ionic proportions, the latter can be evaluated by dividing the ionic concentration of each constituent with its atomic or molecular weight. Expressing the major elements in terms of their molar concentrations and summing up (Table 2.2), it is evident that the concentration of total (expressed as) cations exceeds that of anions. This difference is expressed as the alkalinity of sea water which gives it a character of weak base, with a pH of 8.0 ± 0.2.

Table 2.1: Average Concentrations of the Major Ions in Sea Water, in Parts per Thousand by Weight (g kg^{-1} or g l^{-1})

Ion	*‰ by Weight*	
Chloride, Cl^-	18.980	Negative ions (anions) total = 21.861‰
Sulphate, SO_4^{2-}	2.649	
Bicarbonate,* HCO_3^-	0.140	
Bromide, Br^-	0.065	
Borate, $H_2BO_3^-$	0.026	
Fluoride, F^-	0.001	
Sodium, Na^+	10.556	Positive ions (cations) total = 12.621‰
Magnesium, Mg^{2+}	1.272	
Calcium, Ca^{2+}	0.400	
Potassium, K^+	0.380	
Strontium, Sr^{2+}	0.013	
Overall total salinity	34.482 ‰	

* Includes carbonate, CO_3^{2-}

(*Source*: Open University, 1995)

If one compares the average percentage by weight of most abundant 10 elements in the earth' s crust (Table 2.3) with those in sea water, there are some distinct deviations. Only four of the elements in Table 2.1 appear in Table 2.3, (Ca, Na, K and Mg) though not in the same sequence. The three most abundant elements (Si, Al and Fe) besides other minor ones (Ti, Mn and P) do not appear at all in Table 2.1. This is due to the differences in the solubility and reactivity of different elements while they are being transported as weathering products in solution into the oceans by rivers. Many of the common elements in rocks, such as Si, Al, Fe etc., are not soluble to a large extent, and hence they are transported and deposited in the ocean mainly as solid particles along with sand and clay. Other elements such as Na, Mg, Ca, K etc. are relatively more soluble and are transported mainly in solution. Hydrothermal solutions associated with sea floor spreading supply some elements to sea water and remove others from it. Further, the relative amounts of dissolved constituents within the oceans are controlled by complex biogeochemical processes which are outlined in the next chapters (3-4). All these factors explain the apparent discrepancies in the abundance of major elements in sea water when compared with that of the earth' s crust.

Table 2.2: Composition of Major Components of Sea Water (19.374‰ Cl; 35‰ S)

	g/kg	*m mol/kg*	*m equv/kg*
Cations			
Sodium	10.77	468.46	468.46
Potassium	0.399	10.20	10.20
Magnesium	1.290	53.07	106.14
Calcium	0.4121	10.28	20.56
Strontium	0.0079	0.09	0.18
Total	-	-	**605.54**
Anions			
Chlorine	19.354	545.90	545.90
Bromine	0.0673	0.84	0.84
Fluorine	0.0013	0.07	0.07
Sulphate	2.712	28.23	56.46
Bicarbonate	0.1385	2.27	-
Borate	0.027	0.4	-
Total	-	-	**603.27**

Major elements in general, have long residence times (10^6- 10^8 years) in sea water (Table 1.2) as they are geochemically inert and hence remain in solution for a long time. On the other hand minor elements have relatively short residence times (10^3–10^4 years) as they are geochemically reactive and hence are removed rapidly from sea water (Table 1.2).

Table 2.3: Approximate Average Percentages by Weight of the Ten Most Abundant Elements (other than oxygen) in the Earth's Crust

Element	*% by Weight*
Silicon, Si	28.2
Aluminium, Al	8.2
Iron, Fe	5.6
Calcium, Ca	4.2
Sodium, Na	2.4
Potassium, K	2.4
Magnesium, Mg	2.0
Titanium, Ti	0.6
Manganese, Mn	0.1
Phosphorus, P	0.1

Source: Open University, 1995.

2.1.1 Constancy of Ionic Composition

The constancy of ionic composition of sea water is an important concept in chemical oceanography and was first propounded by Dittmar (1884) from the results of analysis of the water samples collected by H.M.S. Challenger. It states that "the total amount of major dissolved ions can vary from place to place in the oceans, but their relative proportions with chloride ions remain virtually constant". In other words, the salinity can change from place to place in the oceans, but the ratio of concentration of any particular ion listed in Table 2.1 to the salinity remains virtually constant. This is one important characteristic feature of major and conservative elements (with few exceptions). Salinity of sea water in the oceans depends mostly on the balance between evaporation and precipitation, and the extent of mixing between surface and deep waters. However, changes in salinity, in general, have no effect on the relative proportions of major elements– their concentrations also change in the same proportions in order to keep their ionic ratios constant. The ionic ratios of major elements to chlorinity in several oceans and seas are given in Table 2.4. Evidence available so far

Table 2.4: Major Constituent Concentration-to-Chlorinity Ratios for Various Oceans and Seas

Ocean or Sea	Na / ‰ Cl	Mg / ‰ Cl	K / ‰ Cl	Ca / ‰ Cl	Sr / ‰ Cl	SO_4 / ‰ Cl	Br / ‰ Cl
N. Atlantic	–	–	0.02026	–	–	–	0.00337–0.00341
Atlantic	0.5544–0.5567	0.0667	0.01953–0.0263	0.02122–0.02126	0.000420	0.1393	0.00325–0.0038
N. Pacific	0.5553	0.06632–0.06695	0.02096	0.02154	–	0.1396–0.1397	0.00348
W.Pacific	0.5497–0.5561	0.06627–0.0676	0.02125	0.02058–0.02128	0.000413–0.000420	0.1399	0.0033
Indian	–	–	–	0.02099	0.000445	0.1399	0.0038
Mediterranean	0.5310–0.5528	0.06785	0.02008	–	–	0.1396	0.0034–0.0038
Baltic	0.5536	0.06693	–	0.02156	–	0.1414	0.00316–0.00344
Black	0.55184	–	0.0210	–	–	–	–
Irish	0.5573	–	–	–	–	0.1397	0.0033
Puget Sound	0.5495–0.5562	–	0.0191	–	–	–	–
Siberian	0.5484	–	0.0211	–	–	–	–
Antarctic	–	–	–	0.02120	0.000467	–	0.00347
Tokyo Bay	–	0.0676	–	0.02130	–	0.1394	–
Barents	–	0.06742	–	0.02085	–	–	–
Arctic	–	–	–	–	0.000424	–	–
Red	–	–	–	–	–	0.1395	0.0043
Japan	–	–	–	–	–	–	0.00327–0.00347
Bering	–	–	–	–	–	–	0.00341
Adriatic	–	–	–	–	–	–	0.00341

Source: Horne, 1969.

indicates that seven out of eleven (Table 2.1) major constituents (Cl^-, Br^-, SO_4^{2-}, Boron, Na^+, K^+, Sr^{2+}) maintain constant ratio of their ionic concentrations to salinity (within the analytical error) except in enclosed seas such as Baltic and Black Seas. Hence, they are designated as conservative elements, while others (minor elements) non-conservative.

However, few exceptions to this generalization have been observed in certain seas with regard to Ca^{2+}, HCO_3^- and Mg^{2+} ions. Calcium is enriched relative to salinity in deep and intermediate layers by about 0.5 per cent when compared with surface. This is attributed to biological involvement of both Ca^{2+} and HCO_3^- ions. Surface sea water is generally supersaturated with respect to calcium carbonate minerals, and the organisms living in the water rapidly abstract Ca^{2+} along with the CO_3^{2-} or HCO_3^- to form their skeletal frame work. When these organisms die they sink and their skeletons begin to dissolve, thus releasing CO_3^{2-} (or HCO_3^-) and Ca^{2+} ions to the sea water. This is the most likely mechanism for the observed high Ca: S and HCO_3: S ratio in intermediate and deep waters. Mg^{2+} can substitute for Ca^{2+} in the skeletons of marine organisms. In addition, Ca and Mg are important participants in the reactions between hot sea water (hot brines) and crustal rocks in sea floor hydrothermal systems which may result in the change of their ionic ratios with salinity.

It is surprising to know that although most ocean (surface) waters are supersaturated with $CaCO_3$, its precipitation does not occur frequently leading to the lowering of Ca:S ratio in surface waters. This is due to the inhibiting effect of Mg^{2+} ions in solution, since much of the CO_3^{2-} ion (67 per cent of total) is present as ion pairs (Mg^{2+}:CO_3^{2-}), thus preventing the availability of CO_3^{2-} ion of organisms to precipitate $CaCO_3$. Calcareous skeletal material is made up of either calcite or aragonite, both of which have the same chemical formula ($CaCO_3$) but different crystalline structures. However, the aragonite dissolves more readily in sea water than calcite since aragonite structure is less stable (thermodynamically) than calcite.

2.1.2 Variation of Ionic Ratios of Major Elements Due to Local Conditions

In certain atypical marine environments, away from the open oceans, the major elements exhibit large variations in their ionic ratios with salinity. Such regions include:

1. Semienclosed seas (such as Baltic and Black Seas), estuaries and other regions where there is substantial inflow of river water, which not only contains significantly less dissolved salts than sea water, but also have very different ionic ratios.
2. Anoxic basins and Fjords (Black Sea, Swedish and Norwegian Fjords), where the bottom circulation is severely restricted, for example, by the presence of a sill (a subsurface barrier) at the mouth of the basin preventing the full exchange of bottom and surface waters. In such cases, the bacterial breakdown (oxidation) of organic matter in the bottom water leads to depletion of dissolved oxygen, thus leading to anoxic conditions. Under these conditions, the bacteria abstracts oxygen from SO_4^{2-} ions leading to the formation of sulphide (S^{2-}) ion, a part of which may be precipitated as iron sulphide (FeS_2). This results in the removal of SO_4^{2-} ions from sea water and thus decreasing the SO_4^{2-}:S ratio with respect to normal sea water.
3. Warm, shallow water regions such as Bahama Bank which are characterized by active chemical/biological precipitation of calcium carbonate, leading to the changes in the Ca: S ratio.
4. Areas of sea floor where interstitial or pore waters in the sediments participate in a wide variety of reactions with the sediment particles during compaction, after the formation of sedimentary layer (diagenesis). Such reactions lead to considerable changes in ionic ratios.
5. Regions of sea floor spreading and active submarine volcanism where hot sea brine circulates through cracks and fissures in the oceanic crust. Interaction of basalt with hot brine results in removal of Na^+, Cl^- and Mg^{2+} from sea water and addition of Ca^{2+} leading to lowering of Na:S, Mg:S and increasing Ca: S, in comparison with values for normal sea water. A typical example is Red Sea brines which have salinities 7-9 times higher (225-326 psu) that of normal sea water, and temperature in the order of 50° C.

2.2.0 Minor Elements

Minor constituents make up about 0.1 per cent of the dissolved salts in the oceans. Although the distinction between the major and minor is somewhat arbitrary, the latter are defined as those with concentrations of about 1 part per billion (1: 10^9) by weight, or less. On this basis elements below titanium (Table 1.1) are categorized as minor elements.

Apart from this, another distinction between major and minor constituents is their conservative and non-conservative behaviour. While oceanic distribution in general, of individual major elements closely relate to that of salinity because of constancy of their ionic composition, the trace constituents on the other hand, generally behave non-conservatively being effected by biological and chemical processes in which they are added to or removed from sea water. Further, the residence times (Table 1.2) of major elements are relatively high (10^6-10^8 years) because they are in general, geochemically unreactive, while the residence times of minor constituents are lower (10^3-10^4 years) since they are geochemically more reactive in sea water.

Minor elements such as copper, lead, zinc, manganese, cadmium, mercury, nickel, etc. are important since they are introduced into the oceans by anthropogenic inputs apart from natural sources like rivers and winds.

2.2.1 Biological Controls on the Concentration of Trace Metals in Sea Water

One of the reasons why minor elements are more reactive than major elements in sea water is their direct or indirect reaction with organisms. It is believed that for every non-conservative element there is at least one organism which can concentrate and get enriched by the element. Shell fish have long been known to concentrate trace metals, with enrichment factors of many thousands (Table 2.5). Similarly plankton concentrate trace elements more strongly than do organisms further along the food chains. In general, lower organisms concentrate trace metals more strongly than higher organisms. However, the enrichment mechanisms are quite complex. Some of the metals concentrated by the organisms, appear to be vital for growth (bio-limiting), whereas the uptake of others seems to serve no apparent useful purpose (bio-unlimiting). The ability to concentrate a trace metal from sea water depends both on the nature of the organism and the metal. For example:

1. Copper is incorporated in haemocyanin, the blood pigment of molluscs and crustaceans.
2. Vanadium is concentrated by certain ascidians or sea-squirts (*e.g. Ascidea ceratoides*) that form an organo-vanadium complex of blood pigment, whereas other species concentrate niobium to a larger extent than vanadium.

Table 2.5: Enrichment Factors* for the Trace Element Composition of Selfish Compared with the Marine Environment

Element	*Enrichment Factors*		
	Scallop	*Oyster*	*Mussel*
Ag	2300	18700	300
Cd	2260000	318700	100000
Cr	200000	60000	320000
Cu	3000	13700	3000
Fe	291500	68200	196000
Mn	55500	4000	60
Mo	90	30	60
Ni	12000	4000	14000
Pb	5300	3300	4000
V	4500	1500	2500
Zn	28000	110300	9100

$$\text{*Enrichment Factor} = \frac{\text{Weight of element per unit weight of organism}}{\text{Weight of element per unit weight of sea water}}$$

Source: Open University, 1989.

3. Some sponges (*e.g. Dysidea crawshayi*) accumulate titanium whereas others of the same genus (*e.g. Dysidea etheria*) do not.
4. The order of affinity of cations for organisms is 4+ and 3+ (oxidation state) elements >2+ transition elements > 2+ group IIA metals >1+ group I metals. For example, the

order of affinity of plankton is shown as Fe > Al > Ti > Cr; Si > Ga; Zn > Pb > Cu > Mn > Co > Ni > Cd. In general, the heavier elements in a particular group of the periodic table are taken up more strongly than the lighter ones.

5. The affinity for anions increase with increasing ionic charge, and in related species with increasing atomic weight of the central atom. For example, the affinity increases in the order, $I^- > Br^- > Cl^-$; $WO_4^{2-} > MoO_4^{2-} > SO_4^{2-}$.

Several mechanisms which occur either in isolation or collectively, have been suggested for the enrichment of trace metals in organisms. These include:

1. Ingestion of particulate suspended matter such as clays or organic particulates which have scavenged minor or trace elements from sea water. This mechanism is most significant for filter-feeding organisms.
2. Ingestion of elements already concentrated in food material. For example, plankton concentrate the trace metals, and organism higher in the food chain consume the plankton. Such progressive concentration has been observed in Hg, DDT and PCB' s (polychlorinated biphenyls).
3. Complexing of metals with organic molecules. At the mucus surface of digestive glands or gills of many organisms there are large organic molecules such as glycoproteins, which can form complexes with metal ions.
4. Incorporation of metal ions into physiologically important systems (*e.g.* copper into haemocyanin, and cobalt into vitamin B_{12}).

Biological processes, thus, clearly control the trace element composition of sea water. However, direct evidence of their biolimitation is less readily available because of their very low concentration (ppb or less) levels in sea water. Two typical examples of covariation of Cu and Ni with nutrients as illustrated in Figures 2.1a and 2.1b indicate their involvement in biological processes.

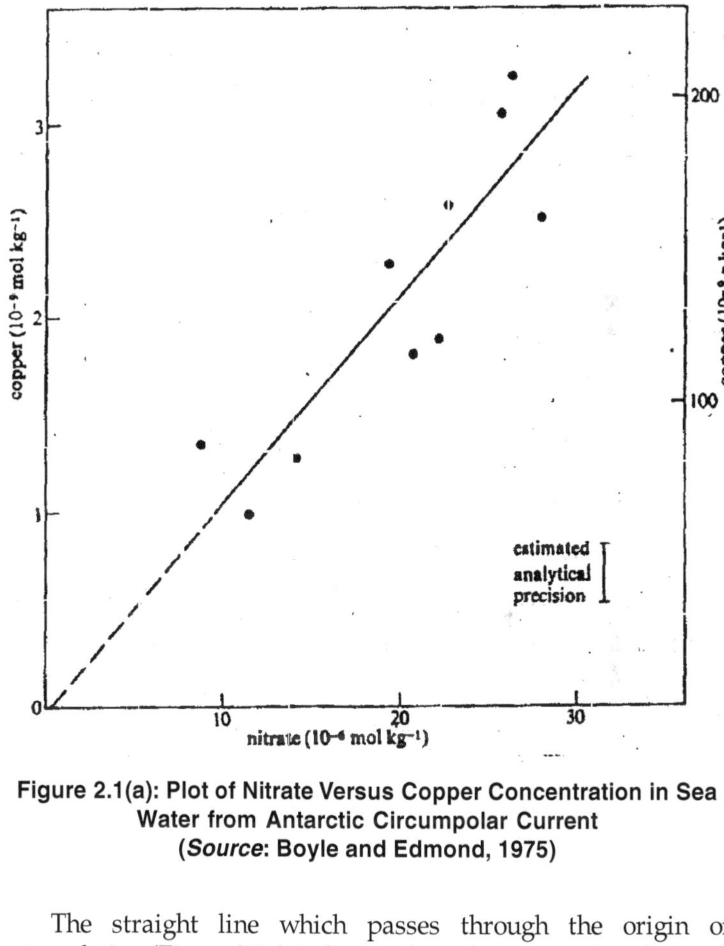

Figure 2.1(a): Plot of Nitrate Versus Copper Concentration in Sea Water from Antarctic Circumpolar Current (*Source*: Boyle and Edmond, 1975)

The straight line which passes through the origin on extrapolation (Figure 2.1a) indicates that when nitrate is exhausted, copper is also exhausted in Antarctic waters indicating that copper involves in biological activity and thus appears as biolimiting. The nickel profiles (Figure 2.1b) have similarities for both phosphate and silicate, which are found in the soft and hard parts of organisms respectively. We can, therefore, infer that nickel is biologically controlled, and that it may be associated with both soft organic tissue as well as hard skeletal materials.

Trace elements concentrated in the soft parts of organisms will generally be redissolved in water near the surface when the

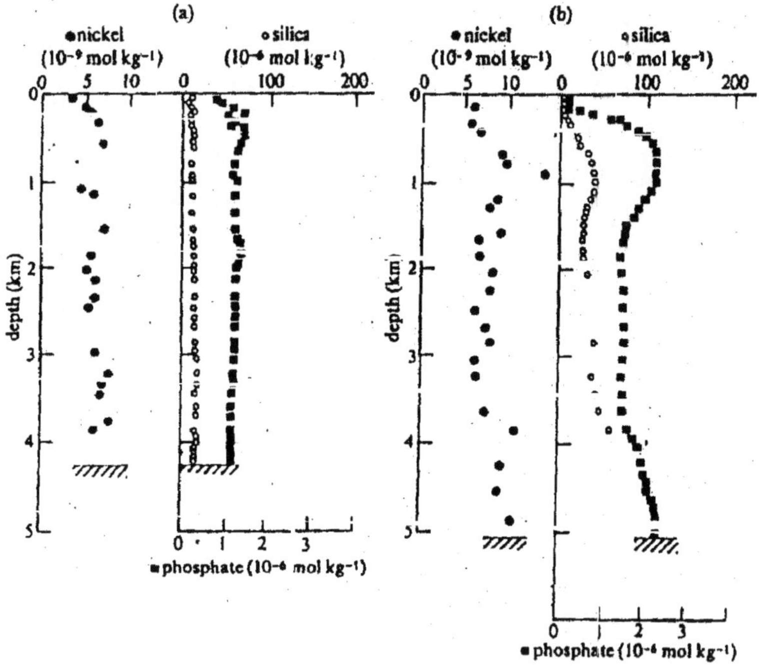

Figure 2.1(b) Concentration-depth Profiles of Nickel, Phosphate and Silicate in (a) the North Atlantic and (b) the tropical Atlantic (*Source*: Sclater *et al.*, 1976)

organisms die and decompose. In this case, the organic tissue provides only a short term sink for trace elements. On the other hand, trace elements concentrated in skeletal material will partly be returned to deeper waters by redissolution, and partly preserved in sediments (opal, carbonate etc.), thus providing a long term sink.

Trace elements incorporated either in the organic matter of soft tissues or into skeletal material will more or less get depleted in surface waters as in the case of nutrients (NO_3^-, PO_4^{3-} and SiO_4^{4-}) and enriched in the deep ocean. They are classified as recycled elements, most of them have residence times of less than 10^6 years. Dissolved constituents in this group can be further classified as biolimiting and bio intermediate, according to whether or not their availability in surface waters can limit primary production as described in 2.2.3.

The redox state as well as the speciation of trace elements affects their behaviour in marine biological systems. For example, selenium is more readily available to marine organisms in its hexavalent form (SeO_4^{2-}) than in the tetravalent form (SeO_3^{2-}). Similarly trivalent arsenic (AsO_3^{3-}) is more toxic than the pentavalent form (AsO_4^{3-}). Mercury is more readily taken up by marine organisms in the form of organo mercury compounds (methyl and dimethyl mercury) than in its simple inorganic ionic forms (Hg^+ or Hg^{2+}). Similarly marine organisms prefer lead in the organo form (diethyl lead) than in its simple ionic forms (Pb^{2+} or Pb^{4+}).

2.2.2 Inorganic Controls on the Concentration of Trace Metals in Sea Water

As organic matter does not accumulate to significant extent in most deep water layers, this process is unlikely to be the ultimate control of trace elements in sea water. The most obvious inorganic control is solubility. This is evident from the data on measured solubility of the least soluble compounds likely to be formed of some trace elements in sea water (Table 2.6).

Table 2.6: Potential Solubility Controls on the Concentrations of Trace Elements in Sea water

Element	*Least Soluble Compound Under Sea water Conditions*	*Saturation Concentration Under Sea water Conditions (mol.l^{-1})*	*Upper Limit of Observed Concentration (mol.l^{-1})*
Lanthanum	$LaPO_4$	$8x10^{-12}$	$2x12^{-11}$
Thorium	$Th_3(PO_4)_4$	$2x10^{-12}$	$2x10^{-13}$
Cobalt	$CoCO_3$	$3x10^{-7}$	$6x10^{-9}$
Nickel	$Ni(OH)_2$	$6x10^{-4}$	10^{-7}
Copper	$Cu(OH)_2$	$2x10^{-6}$	$5x10^{-8}$
Silver	AgCl	$6x10^{-5}$	$3x10^{-9}$
Zinc	$ZnCO_3$	$2x10^{-4}$	$2x10^{-7}$
Cadmium	$CdCO_3$	10^{-5}	10^{-9}
Mercury	$Hg(OH)_2$	10^{-8}	$8x10^{-10}$
Lead	$PbCO_3$	$3x10^{-6}$	$2x10^{-10}$

Source: Open University, 1989.

By comparing third and fourth columns of Table 2.6, it is evident that except (La), all other elements have smaller concentrations in the fourth column than the third column, indicating that they are undersaturated in sea water. On the other hand, their short residence times (Chapter 1.5.0) suggest that they are removed quickly from the sea water–they move rapidly from source to sink. How do we account for this contrasting behaviour? Some *plausible* mechanisms of trace element removal from sea water are given below:

1. *Formation of discreet solid phases*: Minor or trace elements such as Co or Pb may form compounds such as [(Fe, Co) OOH] along with geothite or [(Mn, Pb) O_2] along with manganese oxide unlike the conventional precipitates such as $CoCO_3$ or $PbCO_3$ as shown in Table 2.6. This results in their removal from solution at concentrations less than the saturated concentrations (Table 2.6).
2. *Adsorption*: Adsorption of metal ions on the surfaces of particles of both organic (detritus) and inorganic (clay minerals, hydroxides) origin, as they sink in the water column, removes them from sea water. One of the best example is the adsorption of Cu, Ni, Co, Pb, Zn, Cd etc. on the hydrous oxides of iron and manganese during the formation of ferro-manganese nodules. Elements adsorbed on the small particles are removed from the water column as large particles capture them and carry downwards. This process called scavenging is extremely efficient for removal of trace elements from sea water. The residence times of scavenging particles will be less than 10^3 years, and for many it is less than 100 years.
3. *Oxidation-reduction equilibria*: A change in the valency state of an element upon oxidation or reduction can greatly affect its solubility. Many elements (Fe, Cr, V) in sea water are able to participate in oxidation-reduction (redox) reactions because they have variable valency. The degree of redox conditions of sea water will control the solubility equilibria of the above elements. Iron in trivalent state (Fe III) is much less soluble than in divalent state (Fe II) and hence under oxidising conditions of sea water, iron will be less soluble and present as either $Fe(OH)_3$ or FeOOH (goethite). However, under reducing conditions, iron will be more soluble as it will be present in sea water in the form of Fe^{2+}.

Two other examples are cobalt and manganese, which are present in sea water as Co^{2+} and Mn^{2+} respectively under reducing conditions, both can be oxidised to less soluble Co^{3+} and Mn^{4+} and precipitate as hydroxides or hydrated oxides under oxidising conditions. Hence a knowledge of the redox state of the multivalent elements in solution is important for predicting the rate and effects of chemicals introduced by man into the marine environment. We need to know whether they tend to dissolve or precipitate via solution equilibria controlled by the redox equilibria.

2.2.3 Classification of elements in sea water

Elements present in sea water can be broadly classified into three categories based on concentration-depth profiles.

1. Conservative constituents (some times called accumulated elements) which interact only weakly with the biological particle cycle, and (nearly all) have residence times of 10^6 years or more. They include major constituents discussed in 2.1.0, as well as several trace elements (B, Br, Cl, Cs. F, K, Li, Mg, Mo, Na, Rb, S, Tl, U, Ag, As, Ba, Be, C, Ca, Cd and Cr). Some times they are also referred as bio unlimited constituents.
2. Recycled elements (biolimiting or bio intermediate) which are incorporated in the organic matter of soft tissues or into skeletal material, interact strongly with biological cycle and have residence times less than 10^6 years. They include Cu, Dy, Er, Eu, Fe, Gd, Ge, Ho, I, La, Lu, N, Nd, Ni, P, Pd, Pr, Pt, Ra, Sc, Se,Si, Sm, Sr, Tb, Tm, V, Yb and Zn, and.
3. Scavenged elements which are adsorbed on small particles and removed from the water column by bigger particulates. They have residence times less than 10^3 years. They include Al, Bi, Ce, Co, Hg, Mn, Pb, Sn, Te and Th.

Vertical profiles of all the three categories are shown in Figure 2.2. It is important to note that an element which exhibits a typical recycled profile can also be scavenged from solution (*e.g.* Ni, V, Cu, Zn, Fe). Conversely, an element which exhibits a typical scavenged profile can also involve in the metabolism of organisms (*e.g.* Mn, Co, Al). The concentration–depth profiles merely indicate which of the two pattern of behaviour dominates in sea water.

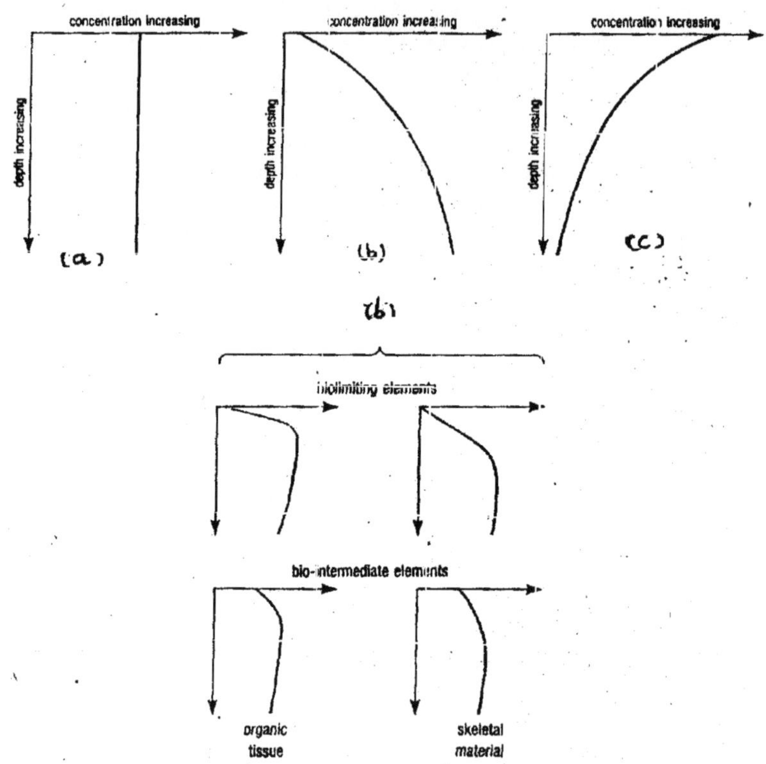

Figure 2.2: Classification of Elements in Sea Water According to Concentration-depth Profiles. (a) Conservative (accumulated) or bio-unlimited elements; (b) Recycled elements and (c) Scavenged elements (*Source*: Open University, 1995a)

2.2.4 Distribution of Minor Elements in Anoxic Environments

Reducing (anoxic or anaerobic) conditions can occur in the oceans only where the rate of consumption of oxygen exceeds the rate of supply. The supply of oxygen below the photic zone depends mainly on advection of surface waters. In isolated basins such as the Black Sea, density stratification of the water and the topographic barriers to mixing (*e.g.* sills) limit the supply of oxygen to deep water

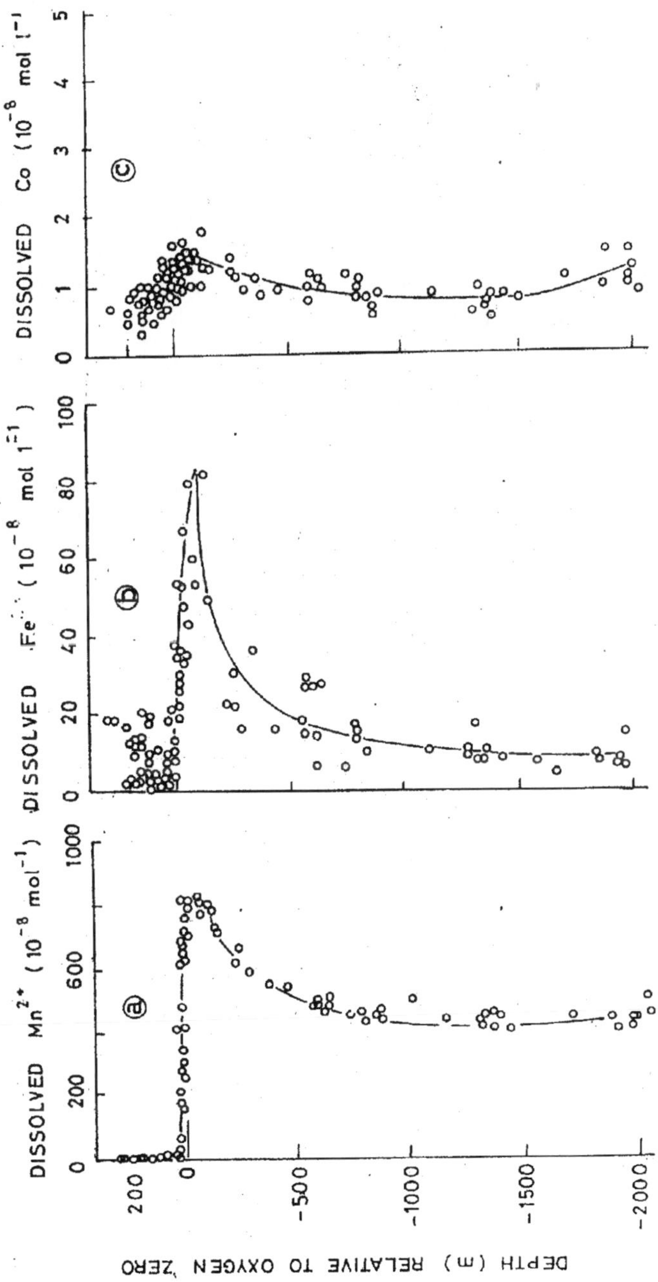

Figure 2.3: Concentration-depth Profiles for Dissolved (a) Manganese, (b) Iron and (c) Cobalt in Black Sea Water. The depth is relative to oxygen zero. (*Source:* Brewer, 1975)

by advection, and anoxic conditions thus prevail. Figures 2.3 (a-c) represents the vertical profiles of dissolved Mn, Fe and Co in the Black Sea. As per the earlier discussion (2.2.2) they should enrich in anoxic and deplete in oxic waters. On the whole, the profiles fairly agree with this assumption. They show sharp increase in the concentrations of Mn, Fe and Co at the oxic-anoxic boundary. At greater depths, however, they decrease again indicating that some other phase is precipitating these elements.

Water in the Black Sea below 200 m depth contains sulphide (S^{2-}) ions as a result of oxidation of organic matter by SO_4^{2-} ions represented by the equation:

$$CH_2O + 1/2\ SO_4^{2-} \longrightarrow 1/2\ H_2S + HCO_3^- \qquad 2.1$$

Since FeS and CoS are insoluble phases, both Fe and Co are removed from Black Sea as their sulphides. Manganese sulphide (MnS), however, is relatively soluble and sulphide formation cannot therefore account for the depletion of Mn in deep waters (Figure 2.3a). It seems likely that in anoxic water the concentration of Mn is controlled by the formation of carbonate $MnCO_3$, rather than the sulphide, MnS.

Multiple Choice Questions:

1. In anoxic basins the ionic ratio of SO_4 to salinity is __________when compared with its normal value in oceans.

   (a) Higher (b) lower

   (c) Remains the same (d) None of the above ( )

2. Major elements in general, have ________residence times in sea water than minor elements.

   (a) Lower (b) Higher

   (c) Equal (d) None of the above

   ( )

3. Calcium is enriched relative to salinity in bottom and intermediate layer of oceans due to

   (a) Inorganic precipitation

   (b) Incorporation in sediments as calcium carbonate

(c) Biological uptake and regeneration
(d) none of the above ( )

4. Minor elements have residence times, in general _______than the major elements in sea water
   (a) Lower (b) Higher
   (c) Similar (d) None of the above
   ( )

5. Minor elements in sea water will have concentrations in
   (a) Parts per thousand
   (b) Parts per million
   (c) Parts per billion and lower
   (d) None of the above ( )

6. Minor elements in sea water behave
   (a) Non conservatively
   (b) Conservatively
   (c) Semi conservatively
   (d) None of the above ( )

7. Enrichment factor is defined as
   (a) $\frac{\text{Weight of the element in organism}}{\text{Weight of the element in sea water}}$
   (b) $\frac{\text{Weight of the element in sea water}}{\text{Weight of the element in organism}}$
   (c) $\frac{\text{Weight of the element per unit weight of sea water}}{\text{Space of the element per unit weight of organism}}$
   (d) $\frac{\text{Weight of the element per unit weight of organism}}{\text{Weight of the element per unit weight of sea water}}$
   ( )

8. The ability of trace metal to concentrate in organisms depend on
   (a) Nature of the metal
   (b) Nature of the organism
   (c) Nature of the metal and the organism
   (d) None of the above ( )

9. A typical example of an anoxic basin is
   (a) Red Sea (b) Black Sea
   (c) Mediterranean Sea (d) Dead Sea ( )

10. Which of the following statement is false
    (a) Plankton concentrate trace elements more strongly than higher organisms.
    (b) The affinity of an ion for organism increases with increasing charge.
    (c) Trace elements concentrated in soft body parts of the organism provide long term sink.
    (d) The redox state as well as the speciation of trace elements affects their uptake by marine organism. ( )

11. What happens when the fixed nitrogen that enters the oceans from river inflow do not match with the amount removed to the sediment?
    (a) Its concentration in oceans decrease.
    (b) Its concentration in oceans increase.
    (c) Its concentration in oceans may increase or decrease.
    (d) No change occurs ( )

Short Answer Questions:

1. Why does sea water not carry a net negative charge even though the proportion by weight of anions (21.861%) far exceeds that of cations (12.621%)?

2. Why the most abundant elements (Si, Al, Fe) in the earth' s crust are present in sea water only in small concentration?

3. What is the ionic ratio of potassium to chlorinity? (concentration of K^+ in sea water = 0.38‰ and chlorinity 18.98‰).

4. Based on the average concentrations in sea water (Na^+ = 10.6‰, Cl^- = 19‰) find out the ionic proportions of Na^+ and Cl^- (Atomic mass of Na=23 and Cl = 35.5).

5. Which of the following statements (a-d) are true and which of them are false?

   (a) The relative proportions of major dissolved constituents in sea water are smaller to those in average crustal rocks. ( )

   (b) Salinity can vary from place to place, but the ratio of Na^+ to salinity will nearly remain constant in all oceans. ( )

   (c) The ratio of Ca^{2+} to salinity will fall where there is significant precipitation of calcium carbonate. ( )

   (d) The order of abundance of major elements (Ca, K) remain more or less same in earth' s crust and in sea water. ( )

6. How the constancy of ionic composition of major elements in sea water is maintained when large amounts of materials are introduced annually by rivers and other sources?

7. Why silicon having more concentration (2 ppm) than fluoride (1.3 ppm) is not considered as a major constituent in sea water?

8. Is sea water normally an oxidising or reducing medium? Why is it so?

9. Are the natural form of Fe and Mn in normal sea water more soluble or less soluble? Why?

10. Based on the general distribution pattern of Cu and Ni in oceans, how do you classify them?
11. What are the probable mechanisms for enrichment of trace metals in organisms?
12. Explain some important processes by which trace metals are removed from sea water.
13. In which form the elements selenium, arsenic, mercury and lead are accumulated by organism?

Review Questions

1. What are the major constituents in sea water and describe their important characteristics?
2. Explain the concept of constancy of ionic composition of major elements in sea water. Under what conditions deviations occur to this concept?
3. Describe how biological controls regulate the concentration of trace metals in sea water.
4. Indicate important inorganic controls which regulate the concentration of trace metals in sea water.
5. Explain the distribution of Fe, Mn and Co in the anoxic environments.

Answers to

Multiple Choice Questions

1. b 2. b 3. c 4. a 5. c 6. a 7. d 8. c

9. b 10. c 11. c

Short Answer Questions

3. 0.02

4. Na^+ = 0.46, Cl^- = 0.53

5. a = false; b = true; c = true; d = true

Chapter 3
Dissolved Gases

Gases dissolved in sea water originate from the earth' s atmosphere, from volcanic activity beneath the sea, and from chemical and biological activity occurring in the sea water itself. The earth (lithosphere) is surrounded by two important phases namely atmosphere and the oceans (hydrosphere). Since atmosphere is the major source of gases, we should consider the abundance of gases in it and their transfer into the oceans. The atmosphere is a homogeneous mixture of three major gases namely nitrogen (78 per cent), oxygen (21 per cent) and argon (1 per cent). Carbon dioxide which is also fairly uniformly distributed in the atmosphere accounts for only 0.03 per cent of the total. Other gaseous constituents present in traces include helium (He), neon (Ne), krypton (Kr), xenon (Xe), radon (Rn) and hydrogen. Few unstable gases (methane, carbon monoxide, methyl iodide, nitrous oxide, sulphur dioxide, dimethyl sulphide etc.) produced by biological processes as well as by man' s activities are also present in the atmosphere in trace quantities. Concentrations of atmospheric gases are expressed in per cent volume or weight. They are also measured in partial pressures which is identical with percentage composition by volume.

3.1.0 Solubility of Gases in Sea Water

Solubility of gases in sea water are expressed in several ways such as $ml.l^{-1}$, $mg.l^{-1}$ and $mol.m^{-3}$. For comparative purposes. Solubility is expressed in terms of percentage saturation of the sample with the gas concerned which can be computed from the equation

$$\%\ \text{Saturation} = 100 \times G/G' \qquad 3.1$$

where,

G is the observed concentration of the gas and G′ is its solubility in water at the appropriate insitu temperature and salinity. Solubility in general, depends on temperature, pressure and salinity of sea water. The solubility of a gas decreases with temperature and salinity and increases with pressure. Figure 3.1 represents the solubilities of four gases in sea water at 24°C. It is evident from the figure that the solubility of CO_2 in sea water is many times greater than that of nitrogen and oxygen. This is due to its reactivity in sea water leading to various carbonate and bicarbonate equilibria.

$$CO_2\,(g) + H_2O \quad H_2CO_3\,(aq) \quad H^+(aq) + HCO_3^-(aq)$$

$$H^+_{(aq)} + CO_{3\ (aq)}^{2-} \qquad 3.2$$

It should be noted that the data in Figure 3.1 is based on the assumption of an equilibrium between the atmosphere and oceans at the air-sea interface. This is probably valid to a first approximation for the major gases (N_2, O_2, Ar and CO_2), but not for many others (He, Ne, Kr, Xe etc.) which occur in very small concentrations. The distribution of gases at deeper levels in the oceans is achieved mainly by currents and by turbulent mixing rather than by diffusion. Downward distribution, in general, is slow and takes a long time. Biological activity plays an important role in the distribution of oxygen and CO_2 below the surface.

3.1.1 Exchange of Gases at Air-sea Interface

Processes in the uppermost few hundredth of a milli-metre (sea surface microlayer) control the transfer of gases from atmosphere to oceans and vice versa. This transfer is a dynamic process, and at equilibrium, when the partial pressure of the gas is same in both the media, molecules enter and leave each phase at the same rate. On

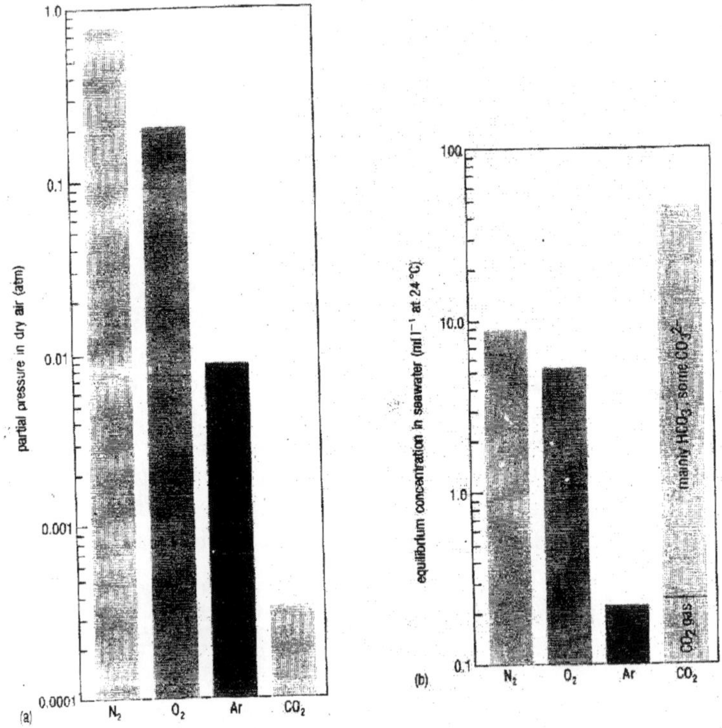

Figure 3.1: (a) Potential Pressure of Four Most Abundant Gases in the Atmosphere; (b) Equilibrium Concentrations in Sea Water (*Source*: Open University, 1995)

the other hand when the partial pressure of a gas in one medium (atmosphere) is higher, there will be net flow of gas from that phase (atmosphere) to the other (sea). The rate of transfer of a gas from the atmosphere to the sea and vice versa is proportional to its partial pressures in the atmosphere (pg) and sea (Pg) respectively. The net rate of accumulation (flux) of gas per unit area of sea surface will be

$$dQ/dt = k_a\, pg - k_s\, Pg \text{ (if } pg > Pg) \qquad 3.3$$

where,

k_a and k_s are the appropriate velocity constants.

At equilibrium dQ/dt = 0, and therefore

$$k_a \, pg = k_s \, Pg$$

The rate of interchange between the gaseous and liquid phase is controlled by the rate of molecular diffusion of the gas through the boundary layer (sea surface microlayer). The thickness of the layer decreases with increasing turbulence of the water and with the wind speed. The diffusion of a gas from atmosphere to the sea through this thin film is given by

$$dQ/dt = A.D\,(pg\text{-}Pg)/dz \qquad 3.4$$

where,

A is the surface area, D is the molecular diffusion coefficient and dz is the thickness of the film.

The flux of various gases (Table 3.1) will vary considerably for a constant partial pressure difference since their solubilities differ widely. Surface sea water is generally supersaturated with N_2O, probably because of bacterial activity and the resulting flux from sea to air is important in oceanic nitrogen budget. The rate at which fixed nitrogen enters the oceans from river inflow and rain is about 8×10^{13} g $N.y^{-1}$. Approximately 10 per cent of this (9×10^{12} g $N.y^{-1}$) is removed to the marine sediments, and the remaining 90 per cent has to be accounted for in order to maintain the steady state composition of the oceans. The sea-air flux of N_2O in Table 3.1 when converted into fixed nitrogen ($1.2 \times 10^{14} \times 28/44$) is equal to 7.6×10^{13} g $N.y^{-1}$ and this makes the balance of the nitrogen input which escapes into the atmosphere.

Carbon monoxide (CO) and methane (CH_4) provide an interesting contrast. Both are produced mainly by organic activity, and their concentrations in surface water are similar. However, their concentration gradients (pg-Pg) across the sea-air interface vary widely because CO is highly oversaturated on surface water while CH_4 is slightly saturated. This is reflected in their annual fluxes ($CO = 4.3 \times 10^{13}$ and $CH_4 = 3.2 \times 10^{12}$ $g.y^{-1}$) differing by about an order of magnitude.

The direction of the flux calculated for methyl iodide (CH_3I) and dimethyl sulphide [$(CH_3)_2S$] is unexpected as these compounds are unstable in the oxygenated environments. They appear to be produced by organic activity near sea surface and exists for long enough to pass into the atmosphere, where they are broken down under more intense ultraviolet radiation. The fluxes of CH_3I account for about 50 per cent while those of $(CH_3)_2S$ account for about 4 per cent for the global budget of iodine and sulphur respectively.

Table 3.1: Worldwide Sea-air Fluxes for Some Gases

Gas	*Total Oceanic Flux (g yr⁻¹)*	*Direction of Net Flux*
SO_2	1.5×10^{14}	air ⟶ sea
N_2O	1.2×10^{14}	sea ⟶ air
CO	4.3×10^{13}	sea ⟶ air
CH_4	3.2×10^{12}	sea ⟶ air
CCl_4	1.4×10^{10}	air ⟶ sea
CCl_3F	5.4×10^{9}	air ⟶ sea
CH_3I	2.7×10^{11}	sea ⟶ air
$(CH_3)_2S$	7.2×10^{12}	sea ⟶ air

Source: Liss and Slater, 1974.

3.2.0 Oxygen

Because of its biological importance and ease with which it can be determined, oxygen is the most extensively studied of the dissolved gases. Its estimation can be carried out on shipboard, immediately, after collection by the most commonly used Winkler' s method or one of its modifications. The sample (sea water) collected in a glass stoppered bottle is treated with a concentrated solution of manganous sulphate (or chloride) and iodide (NaI or KI) in alkaline (NaOH or KOH) solution. The bottle is then carefully stoppered and shaken. Dissolved oxygen reacts with the precipitated manganous hydroxide, oxidising Mn II to Mn IV state under the alkaline conditions. The precipitate is allowed to settle and dissolved in acid (H_2SO_4 or HCl). Iodine liberated during the oxidation of iodide by Mn IV in the acid medium is titrated with standard thiosulphate (hypo) solution using starch as indicator. The following reactions take place in quick succession.

$$Mn^{2+} + 2OH^- \longrightarrow Mn(OH)_2 \qquad 3.5$$

$$Mn(OH)_2 + O \longrightarrow MnO(OH)_2 \qquad 3.6$$

$$MnO(OH)_2 + 4H^+ + 3I^- \longrightarrow Mn^{2+} + I_3^- + 3H_2O \qquad 3.7$$

$$I_3^- + 2S_2O_3^{2-} \longrightarrow 3I^- + S_4O_6^{2-} \qquad 3.8$$

This method gives reliable results with ocean waters, but interference from oxidising (*e.g.* Fe^{3+}, NO_3^-) and reducing (Fe^{2+}, S^{2-}, NO_2^-, organic matter, etc.) agents may cause inaccuracies with polluted estuarine or coastal waters. The interference of Fe^{3+} can be eliminated by using H_3PO_4 instead of H_2SO_4 for acidification, and that of NO_3^- and NO_2^- by adding sodium azide in the Winkler' s alkaline iodide reagent. The interference of reducing agents (Fe^{2+}, S^{2-}, Organic matter) can be prevented by carrying out a preliminary oxidation either by bromine water or by acidic permanganate solution. *In situ* methods for determining dissolved oxygen are available using specific ion electrodes.

Ocean surface waters are consistently supersaturated with oxygen (Figure 3.2) partly due to liberation of oxygen during photosynthesis and partly due to exchange between the air-sea interface. Near the bottom of the euphotic zone, there is a balance between the amount of carbon that phytoplankton fixes by photosynthesis and the amount they dissipate in respiration. The depth at which this balance occurs is called the compensation depth. It is also defined as the depth at which the amount of oxygen produced by phytoplankton during photosynthesis equals the amount they consume in respiration over a 24 hr period as per the reaction which reaches a balance.

$$CO_2\,(g) + H_2O \underset{\text{(metabolic energy)}}{\overset{\text{(light energy)}}{\underset{\text{Respiration}}{\overset{\text{Photosynthesis}}{\rightleftharpoons}}}} (CH_2O) + O_2\,(g) \qquad 3.9$$

Photosynthesis does not cease at the compensation depth, but below it there can be no net plankton growth because more oxygen is being used in plant respiration than is produced by photosynthesis.

3.2.1 Distribution of Dissolved Oxygen

Vertical and horizontal distribution of dissolved oxygen in the oceans results from the interplay of physical and biochemical

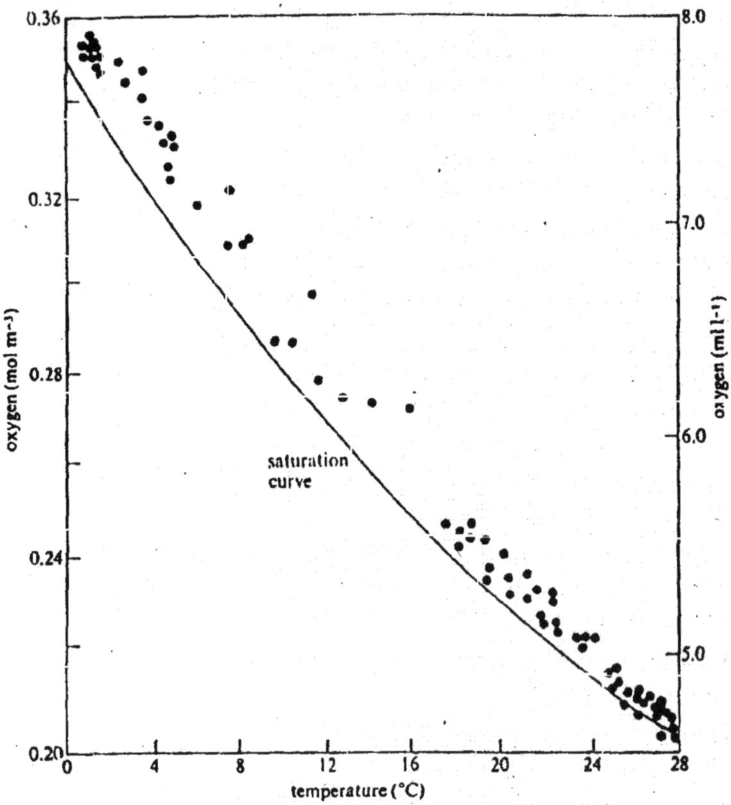

Figure 3.2: The Saturation Curve for Oxygen (Line) and Measured Concentration (Dots) in Atlantic Surface Waters (*Source*: Broecker, 1974)

processes. The vertical distribution which is broadly common for most of the oceans, shows three principal features.

1. A uniform oxygen concentration which is close to the equilibrium value in the surface well mixed layer extending down to the thermocline.
2. Slow or rapid decrease of oxygen concentration with depth leading to a minimum as a result of oxidation of organic matter in the intermediate layers (500– 1000 m depth). In some areas, such as the northern Indian Ocean and eastern

tropical Pacific, the water at these depths is highly oxygen-deficient, and in extreme cases it can become intermittently or completely anoxic (*e.g.* Baltic Sea, Black Sea).

3. Slight increase of oxygen concentration at higher depths (1500–4000 m) because of circulation of cold dense oxygenated water sinking in polar (high latitudes) regions and flowing almost along the ocean bed.

Typical vertical distribution of dissolved oxygen in three regions (Pacific and Atlantic Ocean, and Gulfstream) depicted in Figure 3.3 are in accordance with the well established features.

3.2.2 Distribution of dissolved oxygen in the Indian Ocean

Distribution of dissolved oxygen in the north Indian Ocean is shown in Figures 3.4–3.6. Its spatial distribution at surface and 300 m depth are shown in Figures 3.4 and 3.5 respectively. Dissolved oxygen shows a uniform distribution, close to the saturation value at surface in the north Indian Ocean. An oxygen minimum could occasionally be within the surface layer especially during the pre monsoon (February–May) season. The intensity of solar radiation is very high during this period which causes the maximum primary production to occur a few metres below the sea surface. This together with the vertical stability, may result in the observed oxygen maximum. A sharp fall in dissolved oxygen occurs within the permanent thermocline. The strong density gradients prevent any significant exchange of dissolved oxygen from the euphotic zone to the layers below the thermocline, and the horizontal advection is poor due to the semienclosed nature of the region. These features in conjuction with a high rate of supply of organic matter from the surface result in severe depletion of dissolved oxygen below the thermocline throughout the north Indian Ocean. Dissolved oxygen concentration decreases steadily northward at all subsurface layers in both Bay of Bengal but predominantly in the Arabian Sea (Figures 3.4 and 3.5). In northern parts of the Arabian Sea values below 0.2 $ml.l^{-1}$ are frequently observed at intermediate depths (200–800 m). Vertical profiles of dissolved oxygen, in the Arabian Sea (Figure 3.6) exhibit two minima (Rixen and Haake, 1993). The first minimum is located within the thermocline at about 200 m depth, and the deep minimum between 200–1000 m depth particularly

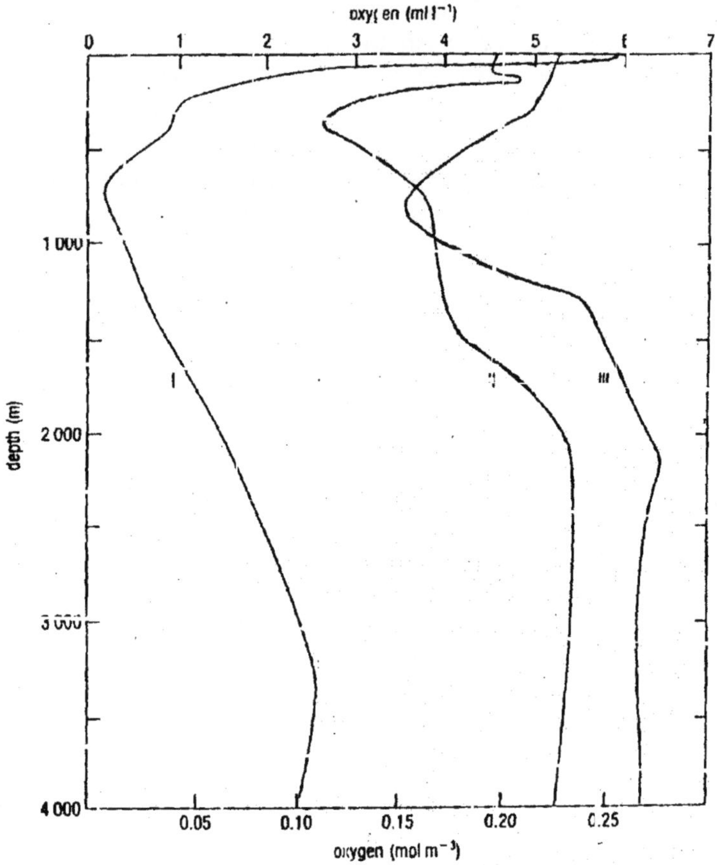

Figure 3.3: The Vertical Distribution of Dissolved Oxygen I: South of California; II: Eastern part of the South Atlantic; and III: Gulf Stream (*Source*: Open University, 1995)

north of the equator. However, in the northwestern Bay of Bengal, only one oxygen minimum with occasional low oxygen concentration (< 0.1 $ml.l^{-1}$) in intermediate layers was observed (Naqvi *et al.*, 1979).

3.3.0 Carbondioxide

Carbondioxide is highly soluble in sea water and enters the ocean from the atmosphere through the air-sea interface. It is also

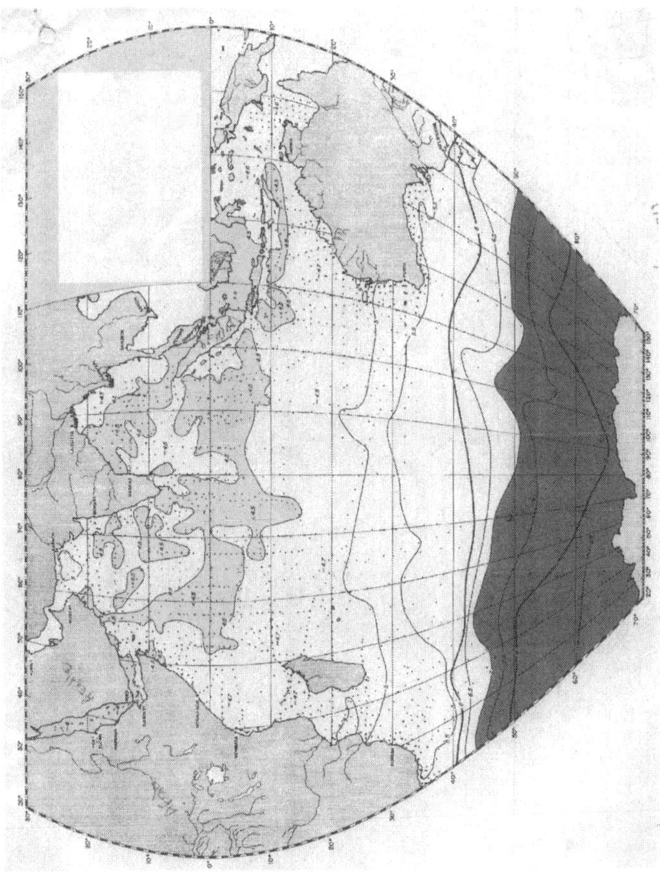

Figure 3.4: Distribution of Dissolved Oxygen (mg.l^{-1}) at Surface in the Indian Ocean (*Source*: Wyrtki, 1971)

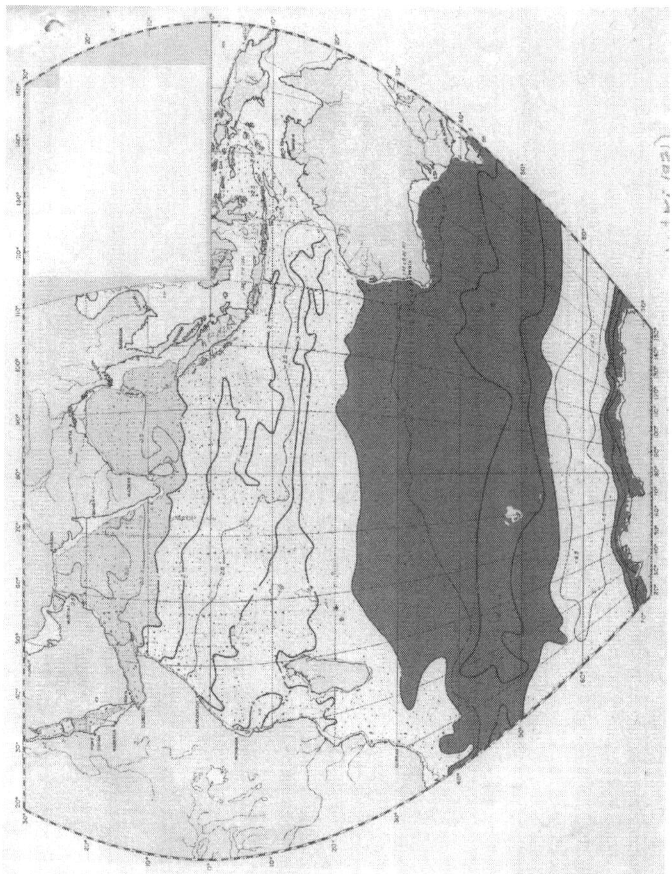

Figure 3.5: Distribution of Dissolved Oxygen (mgl/$^{-1}$) at 300 m in the Indian Ocean (*Source*: Wyrtki, 1971)

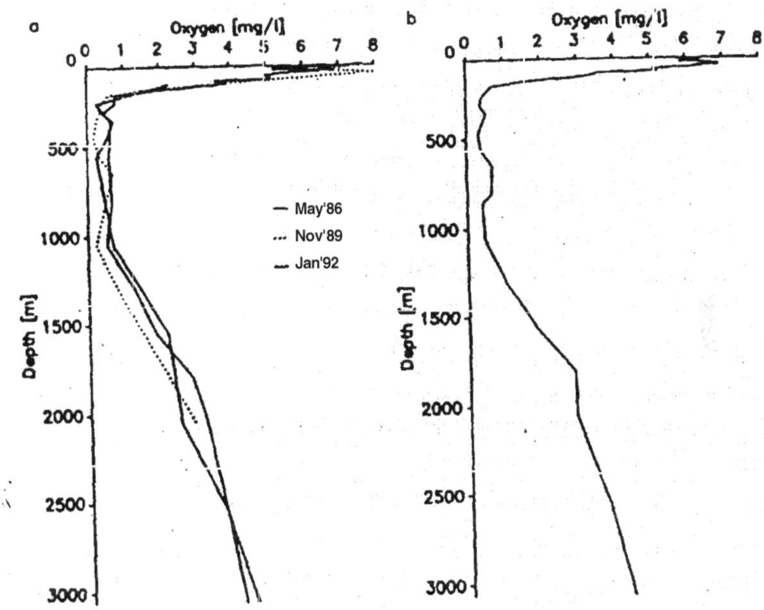

Figure 3.6: Vertical Distribution of Dissolved Oxygen in the Arabian Sea
(a) In three different months and (b) average of three profiles.
(*Source*: Rixen and Haake, 1993)

produced at nearly all depths by the respiration of plants and animals and by the oxidation of organic matter in the sea. It occurs in dissolved form as carbonic acid (H_2CO_3), as carbonate (CO_3^{2-}) and bicarbonate (HCO_3^-) ions, in addition as ion pairs with sodium, calcium and magnesium in small proportions. As such the carbondioxide-carbonate equilibria is the most complicated of all the dissolved gases. A simplified and schematic diagram of carbondioxide-carbonate system is shown in Figure 3.7. The abundance of carbondioxide is controlled by physical properties of sea water (salinity, temperature and pressure), by biological processes (photosynthesis and respiration), and by the formation and destruction of carbonate shells of marine plants and animals. Another significant feature of CO_2^-carbonate equilibria is its capacity to act as buffer and control the pH and alkalinity of sea water and thus provide conducive environment for sustenance of life in the

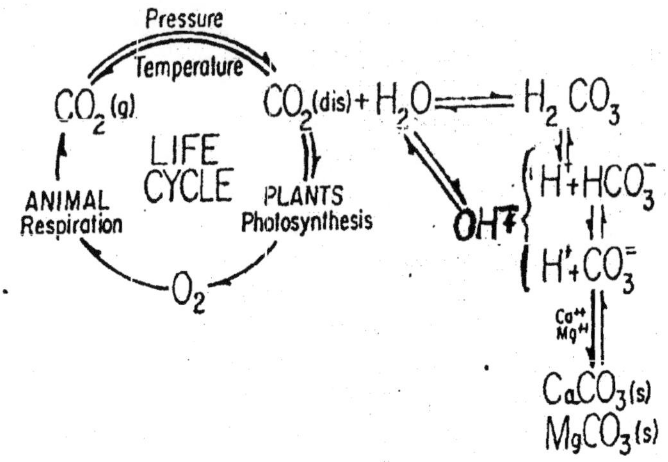

Figure 3.7: The Carbondioxide-carbonate System (*Source*: Horne, 1969)

oceans. In fact, because of the presence of bicarbonate and borate, sea water is said to be the best buffer solution.

3.3.1 Carbondioxide-carbonate Equilibria

When CO_2 gas is equilibrated with water, a series of equilibria are set up as shown below:

$$CO_2\,(g) \rightleftharpoons CO_2\,(aq) \qquad 3.10$$

When the gas diffuses through a thin air-sea boundary layer, it reacts with water to give

$$CO_2\,(aq) + H_2O \rightleftharpoons H_2CO_3\,(aq) \qquad 3.11$$

The forward (hydration) and the reverse (dehydration) reactions are, in general, slow. However, catalytic agents, particularly the enzyme, carbonic anhydrase, can greatly accelerate rates of both the reactions. The H_2CO_3 produced ionizes to

$$H_2CO_3\,(aq) \rightleftharpoons H^+\,(aq) + HCO_3^-\,(aq) \qquad 3.12$$

$$HCO_3^-\,(aq) \rightleftharpoons H^+\,(aq) + CO_3^{2-}\,(aq) \qquad 3.13$$

At pH above 8, as the hydroxyl ion concentration in sea water becomes significant, the following reaction also takes place instantaneously.

$$CO_2\,(g) + OH^-(aq \rightleftharpoons HCO_3^-(aq) \qquad 3.14$$

In aqueous solutions, the balance between various components of the CO_2 equilibria is controlled by the pH as shown in Figure 3.8.

In pure water at pH below 5, almost all the CO_2 is present as dissolved gas. As the pH of pure water is raised, the proportion of bicarbonate ion increases to a maximum of 98 per cent at pH 8.5 and 0° C. Further increase in pH causes decrease of the bicarbonate due to the formation of carbonate ions (eqs. 3.13). With saline waters, the curves for all the three components are displaced towards left (lower pH values). From the diagram (Figure 3.8), we can see at a glance which species is the most predominant in the sea water pH range (8±0.2), clearly they are bicarbonate (HCO_3^-) and carbonate (CO_3^{2-}) ions.

A knowledge of the CO_2-carbonate equilibria is of considerable practical importance. It enables the concentrations of various species

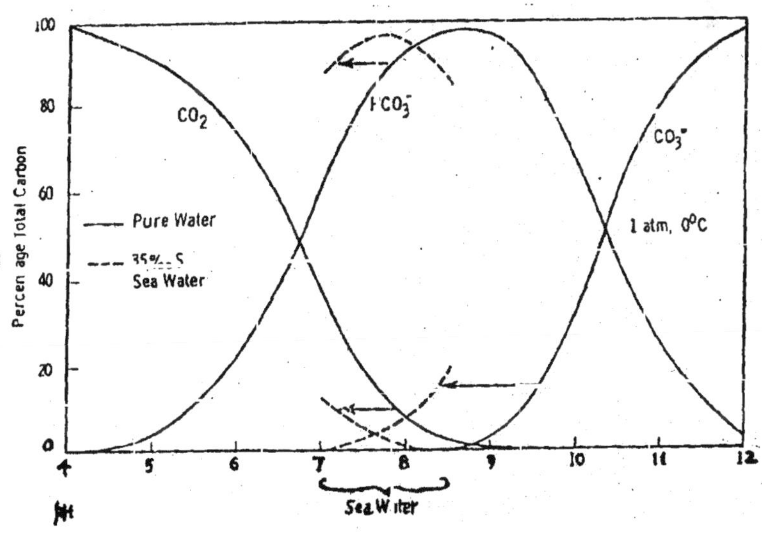

Figure 3.8: Distribution of CO_2- HCO_3^- CO_3^{2-} System in Pure Water and Seawater at 1 atm as a Function of pH (*Source*: Horne, 1969)

such as HCO_3^-, CO_3^{2-}, H_2CO_3 to be calculated from simple measurements such as pH, alkalinity and total carbondioxide CO_2 using the equilibria (eqs. 3.10–3.14) and their equilibrium constants. The anions of weak acids such as HCO_3^-, CO_3^{2-} and borate $[B(OH)_4^-]$ are responsible for the alkalinity of sea water. Hence, total alkalinity (equivalents.l^{-1}) can be quantitatively expressed as the sum of equilibrium concentrations (shown in square brackets) of these ions.

Total Alkalinity =

$$[HCO_3^-] + 2[CO_3^{2-}] + [B(OH)_4^-] + \{[OH^-]-[H^+]\} \qquad 3.15$$

However, in the pH range 7.5-8.5, covering that of normal sea waters, the last term in eq. 3.15 is too small when compared with others and can be neglected. Hence:

$$\text{Total Alkalinity} = [HCO_3^-] + 2[CO_3^{2-}] + [B(OH)_4^-] \qquad 3.16$$

As it has already been mentioned, HCO_3^- and CO_3^{2-} are quantitatively by far the most important species in the sea water pH range. For simplicity the (carbonate) alkalinity (A) is defined as:

$$\text{Alkalinity (A)} = [HCO_3^-] + 2[CO_3^{2-}] \qquad 3.17$$

A simple method to determine alkalinity (A) is to treat an aliquot (100 ml) of sea water with an accurately measured excess (25 ml) of standard 0.01 N hydrochloric acid and to measure the pH of the solution, preferably after expelling CO_2 with a current of air. This pH is the measure of the excess of acid remaining after neutralization of the alkalinity, a knowledge of which enables the calculation of the alkalinity of sea water sample.

Another term called specific alkalinity is often used to facilitate the comparison of alkalinities of various bodies of water (watermasses). It is defined as

$$\text{Specific Alkalinity} = \text{alkalinity} \times 10^3/\text{chlorinity } (‰) \qquad 3.18$$

Different watermasses frequently have characteristic specific alkalinities, these usually have a range of 0.119–0.133 with a mean of 0.126.

Total carbondioxide CO_2, which is sum of the concentrations of all the species of carbondioxide (total dissolved carbon) is expressed as:

$$[\ CO_2] = [CO_2] + [H_2CO_3] + [HCO_3^-] + [CO_3^{2-}] \qquad 3.19$$

However, for simplicity, CO_2 of sea water can be expressed as:

$$[\ CO_2] = [HCO_3^-] + [CO_3^{2-}] \qquad 3.20$$

neglecting the other terms whose proportions are very small at the pH range of sea water.

Total carbondioxide CO_2, is usually estimated by acidifying an aliquot (100 ml) of sea water with excess of HCl, and determining the amount of carbondioxide liberated by infrared absorption spectrometry or gas chromatography techniques. Alternatively CO_2 can be determined by a potentiometric titration of sea water sample with a standard acid.

If we rearrange equation 3.17, we get:

$$A-[CO_3^{2-}] = [HCO_3^-] + [CO_3^{2-}] \qquad 3.21$$

The right hand sides of equations 3.20 and 3.21 are equal. It follows that the left hand sides of these equations should also be equal. Hence:

$$[\ CO_2] = A-[CO_3^{2-}] \qquad 3.22$$

or

$$A-[\ CO_2] = [CO_3^{2-}] \qquad 3.23$$

Hence, by measuring the alkalinity (A) and CO_2, the concentration of the carbonate $[CO_3^{2-}]$ ion can be determined by using eq. 3.23. Substituting this value in equation 3.17 or 3.20 we can obtain the bicarbonate ion $[HCO_3^-]$ concentration. This is of practical importance since the ratio of $[HCO_3^-]$ to $[CO_3^{2-}]$ controls the pH of sea water as shown below.

3.3.2. Control of pH in Sea Water

The pH is defined as:

$$pH = -\log_{10}[H^+] \qquad 3.24$$

where,

$[H^+]$ is the equilibrium hydrogen ion concentration. In sea water, as we have seen, pH is mostly in the range 8.0±0.2 and the variation of pH is controlled chiefly by the components of reactions shown in equations 3.2 and 3.13, namely:

$$HCO_3^-(aq) \rightleftharpoons H^+(aq) + CO_3^{2-}(aq)$$

This reaction is very rapid and the sea water can be assumed to have an equilibrium mixture of these ions. When the above reaction is at equilibrium, we can write:

$$K = \frac{[H^+][CO_3^{2-}]}{[HCO_3^-]} \qquad 3.25$$

where,

K is the equilibrium constant. If we rearrange eq.3.25,

$$[H^+] = K \frac{[HCO_3^-]}{[CO_3^{2-}]} \qquad 3.26$$

Equation 3.26 shows that the ratio of concentrations of HCO_3^- and CO_3^{2-} ions controls the hydrogen ion concentration and hence pH of sea water.

The relationships outlined above have been simplified in order to establish some basic principles of sea water chemistry.

However, we have overlooked the following facts.

1. We have neglected other terms of dissolved carbon CO_2 (eq.3.19) as they are too minute to reckon. But they can not be neglected when accurate measurements and calculations are required.
2. The alkalinity (A) defined (eq.3.17) should (strictly speaking) be called as carbonate alkalinity because other species such as borate ion (eq.3.16) contribute to the total alkalinity.
3. The equilibrium constant of the reaction of bicarbonate-carbonate (eq.3.25) is not strictly constant, but changes with temperature and pressure.

As a general rule, the greater the [CO_2], the smaller the value of [A-(CO_2)] according to eq.3.23 and the greater the value of the $[HCO_3^-]/[CO_3^{2-}]$ ratio and hence the higher the value of $[H^+]$ as per eq.3.26 indicating lower pH of sea water. Conversely, if [CO_2] is lower, the $[H^+]$ value is lower and hence the pH of sea water is higher. For example, one of the few places where inorganic precipitation of $CaCO_3$ occurs is on the Bahama Banks in the Atlantic Ocean, where the sea is shallow and warm, and the salinity is high (greater than 37 psu). The warmer and more saline the water, the lower the solubility of CO_2 with consequent smaller CO_2. Under these conditions the term A-[CO_2] in equation 3.23 will be large leading to higher concentration of CO_3^{2-} ion. It often rises to super saturation level with respect to $CaCO_3$ leading to the precipitation of calcium carbonate in the form of aragonite. So in terms of $CaCO_3$ precipitation, equations 3.17 and 3.26 help to explain why it is likely to precipitate where [CO_2] is low, and why it is likely to dissolve where [CO_2] is high in the oceans. The saturation of calcium carbonate is discussed in more detail under calcium carbonate equilibria (3.3.5).

Table 3.2 summarises the average concentration parameters of CO_2 system (HCO_3^-, CO_3^{2-} and CO_2) for different water bodies. One important finding that emerges from the data given in Table 3.2 is that when total dissolved carbon (CO_2) increases, the concentration of bicarbonate (HCO_3^-) ion also increases while the concentration of carbonate (CO_3^{2-}) ion decreases. This is because the solubility of CO_2 is inversely related to temperature like all other gases. Warm surface waters contain lowest concentration of dissolved carbon, but they have the highest concentration of carbonate ion.

Table 3.2: Carbonate Chemistry of Various Water Types

Water Type	*Bicarbonate* $[HCO_3^-]$ *(mol.m^{-3})*	*Carbonate* $[CO_3^{2-}]$ *(mol.m^{-3})*	*[CO_2]* *(mol.m^{-3})*	*Alkalinity* $A = [HCO_3^-] + 2[CO_3^{2-}]$ *mol.m^{-3}*
Warm Surface	1.65	0.35	2.00	2.35
Cold Surface	1.95	0.20	2.15	2.35
Deep Atlantic	2.10	0.15	2.25	2.40
Deep Pacific	2.35	0.10	2.45	2.55

Source: Broecker, 1974.

3.3.3 Distribution of CO_2

Equilibrium is rarely attained between CO_2 in surface sea water (PCO_2) and in the atmosphere (pCO_2) above it. Physical (surface heating and cooling, and water transport) and biological (photosynthesis and respiration) processes disturb the equilibrium if at all it attains for a short duration. Partial pressure of CO_2 (PCO_2) in sea water is sensitive to total carbondioxide (CO_2), pH, alkalinity and temperature. Hence, the distribution of carbondioxide in sea water exhibit both seasonal and diurnal (variation within a day, 24 hours) changes. During winter (November-February), photosynthesis exceeds respiration and causes CO_2 to fall. However, the winter cooling slightly increases the PCO_2, but this effect is totally compensated by its decrease during photosynthesis. Hence, CO_2 shows a minimum during March-April resulting in its decrease. In summer (April-May), the increase in summer temperature and decay of organic matter tends to raise PCO_2 and this is partially compensated by the fall due to uptake during photosynthesis resulting in a net accumulation and maximum in June-July.

When water sinks below the thermocline, their CO_2 content can be altered only by processes involving advection and eddy diffusion and by biological regeneration. The most important among them are biological regeneration processes which occur as a result of oxidation of organic material (bacterial tissue) falling from the surface. Most of this oxidation occurs in the upper few hundred metres of the water column below the thermocline, reaching maximum at a depth of about 500–1000 metres as shown in Figure 3.9. The depth of CO_2 maximum, in general, coincides with that of nutrient maxima (phosphate and nitrate). The depth profile of CO_2 illustrates the biointermediate (non-conservative) character (Figure 2.2b) in the oceans. At higher depths, its concentration remains more or less constant.

Similarly the depth profiles of CO_2 and phosphate (PO_4-P) in oceans reflect constant carbon:phosphorus ratio in marine organisms. This is due to the fact that when dead organisms fall from the surface layer, they undergo decay and liberate CO_2-C and PO_4-P to the water in the same Redfield (1963) atomic ratio (C:P = 106:1).

3.3.4 Distribution of CO_2 in the North Indian Ocean

Carbondioxide inputs to the Arabian Sea as well as to the Bay of Bengal occurs mainly through the atmosphere and the southern Indian Ocean since the river contribution to these regions is quite meagre. Specific Alkalinity in the northern Indian Ocean was

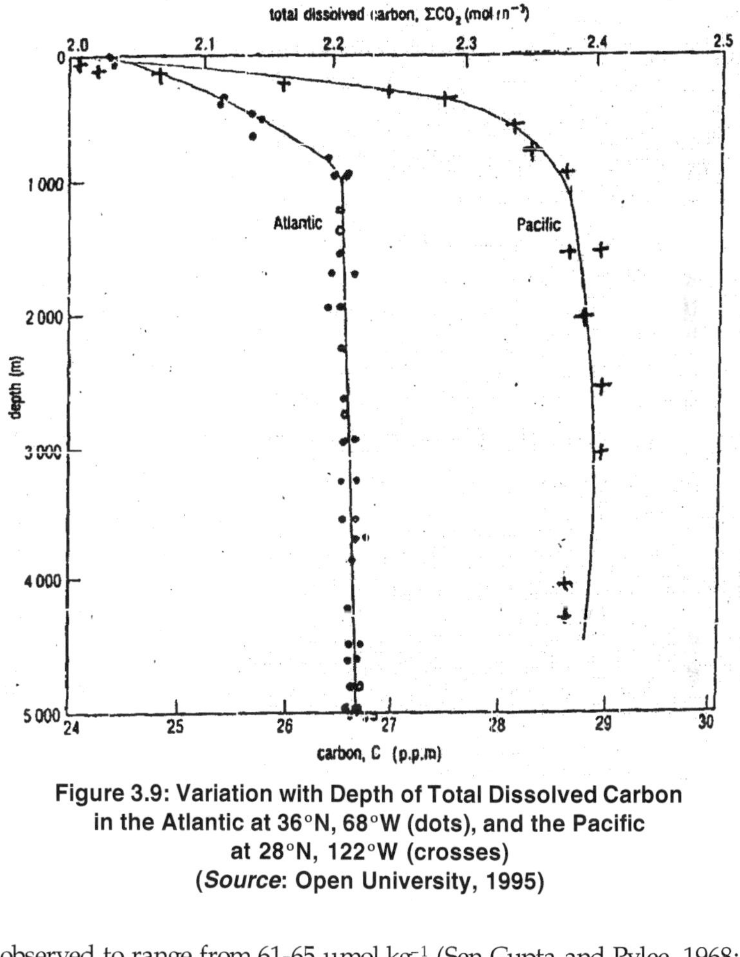

Figure 3.9: Variation with Depth of Total Dissolved Carbon in the Atlantic at 36°N, 68°W (dots), and the Pacific at 28°N, 122°W (crosses) (*Source*: Open University, 1995)

observed to range from 61-65 μmol.kg^{-1} (Sen Gupta and Pylee, 1968; Andersson and Dyrssen, 1994). Geographical variations of CO_2 in the Arabian Sea (Figure 3.10 a) shows significant increase from north to south at all depths. However, this gradient is quite small in the Bay of Bengal (Figure 3.10 b). The northward increase of CO_2 in the Arabian Sea surface waters may be due to higher salinity as well as its supply from subsurface layers through upwelling and vertical mixing. Surface CO_2 in the Bay of Bengal, however, exhibits wide fluctuations probably due to salinity changes.

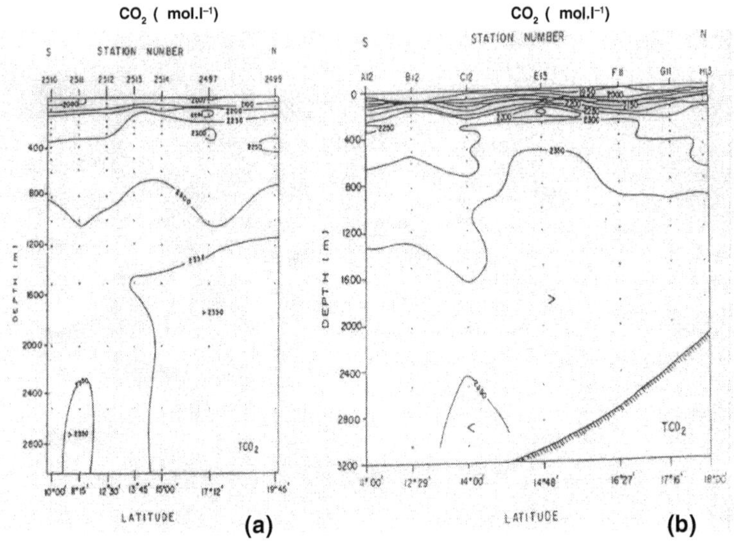

Figure 3.10: North-south Variation of CO_2 ($mol.l^{-1}$) in the Northern Indian Ocean (a) The Arabian Sea and (b) The Bay of Bengal. (*Source*: Dileep Kumar *et al.*, 1995)

The increase of CO_2 (Figure 3.11) with depth in the Arabian Sea reflecting nutrient type behavior is due to the decomposition of organic and inorganic materials leading to its regeneration in the intermediate layers. Surprisingly, inspite of differences in productivity of the Arabian Sea and the Bay of Bengal, and rapid settlement of organic matter in the latter region, the levels of CO_2 are almost the same (Dileep Kumar *et al.*, 1995).

3.3.5 Carbondioxide-calcium carbonate equilibria

So far we have considered the equilibria of carbondioxide and carbonate system relating to air-sea interface. Now we turn our attention to the equilibria involving sea water-sediment (hydrosphere-lithosphere) interface. Important equilibria involved at this interface are

$$CaCO_3 \text{ (s, calcite)} \rightleftharpoons Ca^{2+}\text{(aq)} + CO_3^{2-}\text{(aq)} \quad 3.27$$

$$CaCO_3 \text{ (s, aragonite)} \rightleftharpoons Ca^{2+}\text{(aq)} + CO_3^{2-}\text{(aq)} \quad 3.28$$

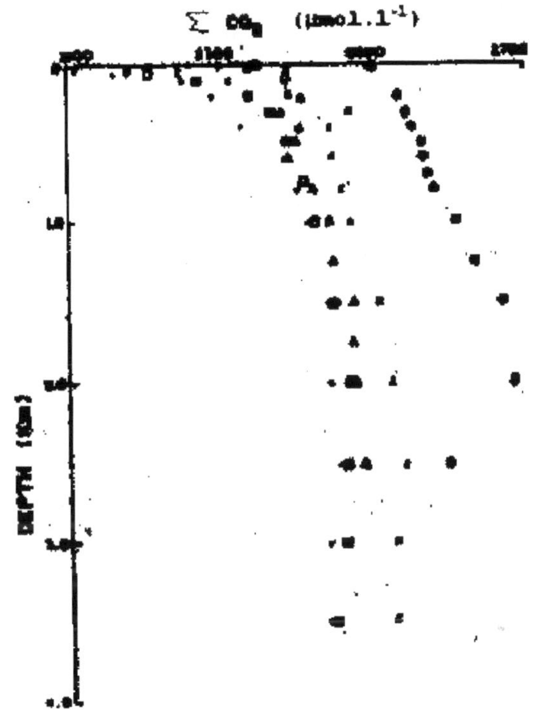

Figure 3.11: Vertical Variation of CO_2 (mol.l^{-1}) in the Arabian Sea at Different Stations (*Source*: Dileep Kumar *et al.*, 1995)

$$CaCO_3\ (s, aragonite) \rightleftharpoons CaCO_3\ (s, calcite) \qquad 3.29$$

$$MgCO_3(s) \rightleftharpoons Mg^{2+}(aq) + CO_3^{2-}(aq) \qquad 3.30$$

Among them, equation 3.30 is not much important in marine sediments. The equilibria involving others are complicated by the existence of two polymorphic forms of $CaCO_3$ namely calcite and aragonite which differ in their crystal structure.

In order to understand the relationship of $CaCO_3$ distribution in sediments and the distribution of Ca^{2+} and CO_3^{2-} ions in sea water, we need to examine the terms saturation, supersaturation and undersaturation more meaningfully. If $CaCO_3$ (s) is placed in contact with sea water and if, the product of concentrations of the ions $[Ca^{2+}]$ and $[CO_3^{2-}]$ during its solution does not reach its solubility product $[(Ca^{2+})(CO_3^{2-})]$, then it is said to be undersaturated. If it reaches its solubility product, it is saturated and if it exceeds the solubility product it is said to be supersaturated. The degree of saturation of $CaCO_3$ in sea water is expressed by the equation:

$$D = \frac{[Ca^{2+}]\,[CO_3^{2-}] \text{ sea water}}{[Ca^{2+}]\,[CO_3^{2-}] \text{ sea water saturated with } CaCO_3} \qquad 3.31$$

Equation 3.31 can be simplified by neglecting the term $[Ca^{2+}]$ which remains virtually constant when compared with high variability of $[CO_3^{2-}]$ in sea water as discussed earlier. Hence:

$$D = \frac{[CO_3^{2-}] \text{ sea water}}{[CO_3^{2-}] \text{ sea water saturated with } CaCO_3} \qquad 3.32$$

when,

$D > 1$		supersaturated		not dissolve
$D = 1$	sea water	saturated	and $CaCO_3$	be in equilibrium
$D < 1$	is	undersaturated	will	tend to dissolve

Apart from temperature and pressure (depth) of sea water, the degree of saturation (D) of $[CO_3^{2-}]$ depends on the polymorphic forms of $CaCO_3$ namely, calcite and aragonite.

Surface waters are always supersaturated with $CaCO_3$ whether it exists as calcite or aragonite (D>1). Organisms forming calcareous skeletons will generate whichever polymorph they find easier to precipitate. For example, many of them precipitate calcite but a few significant proportion (pteropods) precipitate aragonite. Because of their different chemical stabilities, the two forms of $CaCO_3$ show quite different preservation patterns in the deep sea resulting in considerable variation of carbonate compensation depths (CCD) in different oceans.

The depth at which the rate of solution equals or exceeds the rate of supply of the calcareous skeletal remains is called the carbonate compensation depth (CCD). It is also sometimes called as saturation horizon and lysocline. It should be remembered that the calcium carbonate compensation depth is different from that of oxygen compensation depth referred earlier (3.2.0).

Approximate values of the carbonate compensation depth in two oceans (Atlantic and Pacific) are given in Table 3.3. From the data in Table 3.3 it is evident that the compensation depth for calcite is much greater than that for aragonite in both the oceans. This is due to the fact that aragonite is less stable chemically than calcite. Further, the compensation depths for both the minerals are much greater in the Atlantic than in the Pacific Ocean. This is due to the spatial (horizontal) gradient in nutrient concentrations outlined in Chapter 4. Recent studies on the rate of calcite dissolution have indicated that undersaturation in sea water does not become significant until D falls below 0.5. Hence, it should be noted that those hard (skeletal) $CaCO_3$ parts falling where D is less than 0.5 are largely dissolved (destroyed). In sea water where D lies between 0.5 and 1, they are partly dissolved and thus show evidence of considerable destruction. However, in waters where D is greater than 1, the hard parts do not dissolve and hence are preserved. Table 3.4 summerises the saturation concentration of CO_3^{2-} ion in sea water under the range of temperature and pressure conditions likely to be encountered in the world oceans. It points out two very interesting features–Solubility of $CaCO_3$ is higher in cold (2° C) than in hot (24° C) water at one atmosphere pressure (surface). This is opposite to the solubility of the common salts. Secondly, the solubility increases with pressure, so that below the main thermocline (in deep waters) where temperatures are virtually constant (2° C) it is almost entirely a function of pressure. This is because when CO_2 (g) combines with water, the equilibrium shown in equation 3.2 shifts to the right by increased pressure so that the solubility of CO_2 increases with pressure much more than that of most other gases (Le Chatelier' s principle).

3.4.0 Other Minor Gases

Minor gases such as nitrous oxide (N_2O), carbon monoxide (CO), methane (CH_4), methyl iodide (CH_3I) and dimethyl sulphide [$(CH_3)_2S$] are produced by organisms in surface waters in which they are saturated with respect to their atmospheric concentrations. So, for equilibrium to be maintained, they must be transferred into the atmosphere through the air-sea interface. In other words their net flux is from sea to the atmosphere (Table 3.1). However, for sulphur dioxide (SO_2) the net flux is from atmosphere to the sea. Its sources include volcanism, industry (fossil fuel burning and metal smelting), and oxidation of natural organic sulphur compounds. In the atmosphere, it is oxidised to sulphur trioxide (SO_3^-) which quickly combines with water to form sulphuric acid aerosols which contribute to the problem of acid rain. So, much of its input to the oceans may be as sulphate (SO_4^{2-}) ions. The seas around India serve as a source for the greenhouse gases N_2O and CH_4 while they are sinks for CO_2 (Naqvi, 2001).

Table 3.3: Average Levels of the Carbonate Compensation Depth for the Calcite and Aragonite in the Atlantic and the Pacific Oceans

		CALCITE		*ARAGONITE*
			Depth (m)	
			–0	Petropods present in
A		Calcite	–1000	Sediments
T		Unaffected	–2000	(D >1 for aragonite) 2500m
L		(D >1 for calcite)	–3000	
A			–4000	Petropods absent in
N	4500m	————————	–5000	Sediments
T		Calcite dissolves	–6000	(D <1 for aragonite)
I		(D<1 for calcite)		
C				
			Depth (m)	
				Petropods present in
P				Sediments
A			–0	(D>1 for aragonite) 500m
C			–500	
I		Calcite	–1000	
F		Unaffected	–1500	Petropods absent in
I		(D >1 for calcite)	–2000	Sediments
C	3000m	————————	–2500	(D<1 for aragonite)
		Calcite dissolves	–3000	
		(D<1 for calcite)	–3500	

Source: Open University, 1978.

Table 3.4: Carbonate Ion Concentration in Sea water at Saturation for Calcite and Aragonite

Temperature (°C)	*Pressure (atm.)*	*Carbonate ion content at saturation (10^{-6} mol.l^{-1})*	
		Calcite	*Aragonite*
24	1	53	90
2	1	72	110
2	250 (~2500m)	97	144
2	500 (~5000m)	130	190

Source: Broecker, 1974.

Hydrogen sulphide (H_2S) occurs in the land locked basins (seas) which are so poorly ventilated that their waters become completely anoxic. Typical examples are Black Sea, Cariaco Trench and certain Norwegian and Swedish Fjords. Such waters cannot support normal forms of life, and the only living organisms which they contain are specialised forms of bacteria, such as *Disulphovibrio desulphuricum.* These utilise oxygen from SO_4^{2-} ions instead of dissolved oxygen for their metabolic processes. This leads to the liberation of H_2S and setting up of anoxic conditions. Thus, H_2S concentrations as high as 7.3 ml.l^{-1} are encountered at depth in the Black Sea.

Multiple Choice Questions

1. The solubility of a gas in sea water increases with
   (a) Increase of temperature, pressure and salinity
   (b) Increase of temperature and salinity and decrease of pressure
   (c) Increase of temperature, and decrease of pressure and salinity
   (d) Decrease of temperature and salinity and increase of pressure

   ( )

2. The carbonate compensation depth of calcite, in general, is:
   (a) Higher than aragonite
   (b) Lower than aragonite

(c) Similar to the aragonite
(d) None of the above ( )

3. Acid rain is due to the fall of atmospheric gases ______ during precipitation.
(a) Oxides of carbon
(b) Oxides of nitrogen
(c) Oxides of sulphur and nitrogen
(d) None of them ( )

4. Calcium carbonate precipitates from sea water, when its saturation (D) is
(a) Greater than 1 (b) Less than 1
(c) Equal to 1 (d) None of the above ( )

5. The vertical distribution of total dissolved carbon (CO_2) in oceans is similar to
(a) Distribution of dissolved oxygen
(b) Distribution of nutrients
(c) Distribution of pH
(d) Distribution of salinity ( )

6. Warm surface sea waters contain ______ than cold surface waters
(a) Lower CO_2, HCO_3^- and CO_3^{2-} ions.
(b) Lower CO_2 and HCO_3^- but higher CO_3^{2-} ions.
(c) Lower CO_3^{2-} but higher CO_2 and HCO_3^- ions.
(d) Lower HCO_3^- but higher CO_2 and CO_3^{2-} ions. ( )

7. Calcium carbonate concentration at saturation is highest at
(a) Warm surface waters
(b) Cold surface waters
(c) Cold deep waters
(d) Cold intermediate waters ( )

8. Calcium carbonate compensation depth for Pacific Ocean is ________ than Atlantic Ocean.
   (a) Smaller (b) Greater
   (c) Equal (d) None of the above ( )

Short Answer Questions

1. What is the principle condition necessary for the transfer of a gas from
   (a) Atmosphere to the ocean, and
   (b) Ocean to the atmosphere?
2. Why the oceanic fluxes of carbonmonoxide and methane differ eventhough their average concentrations in sea water are almost same?
3. How CO_{2-} carbonate equilibria control the pH of sea water?
4. What is carbonate compensation depth? How does it differs from that of oxygen compensation depth?
5. Why the compensation depth of calcite and aragonite differ from each other?
6. Which of the following Statements (a-e) are true and which of them are false?
   (a) Dissolved carbondioxide and oxygen are conservative constituents in sea water.
   (b) Noble gases are conservative constituents in sea water.
   (c) The oxygen minimum layer is below its compensation depth
   (d) If the partial pressure of a gas in the atmosphere is greater than its partial pressure in the sea, transfer of the gas takes place from atmosphere to the sea.
   (e) Dissolved oxygen exhibits one minimum in the northern Indian Ocean, particularly north of the Equator.

Review Questions

1. Describe the distribution of dissolved oxygen in the northern Indian Ocean.

2. Give an account of the exchange of dissolved gases across the air-sea interface.

3. Explain the carbondioxide-carbonate equilibria effecting the chemistry of sea water at air-sea and sea water-sediment inter face.

4. Define the terms total alkalinity, carbonate alkalinity and specific alkalinity of sea water and explain their significance.

5. Compare the composition of gases in the atmosphere with those dissolved in sea water. How do you account for their differences?

Answers to

Multiple Choice Questions

1. d 2. a 3. c 4. a 5. b 6. c 7. c 8. a

Answer to Short Answer Questions

6. a = False; b = True; c = True; d = True; e = False

Chapter 4
Nutrients

A nutrient element is one which is functionally involved in the processes of living organisms. Certain inorganic elements like nitrogen, phosphorus and silicon in their different chemical forms are essential for the growth and maintenance of life in the sea. In addition, other elements like iron, copper, manganese, zinc, cobalt and molybdenum, and some organic constituents like carbohydrates, amino acids, vitamins etc. are also essential for the growth and sustenance of life in the sea. They are, in general, called as micronutrients. Among them, the most important elements nitrogen, phosphorus and silicon are discussed in detail in this Chapter.

4.0.0 Nutrient Budget in Oceans

Nutrient budget based on the estimates of their inputs and outputs into and out of oceans is given in Table 4.1. Allowing for uncertainties in the available data, the budget for phosphorus in the oceans is apparently balanced. It means that the input from the land and rain balances that lost to the sediments indicating a steady state condition in the oceans. In contrast, the nitrogen budget shows an " imbalance" , namely excess input of about 70 million tonnes. This is rectified by the formation of N_2O or molecular nitrogen in the sea due to denitrification processes. The released nitrogen species will then be transferred to the atmosphere through air-sea interface thus attaining a steady state condition in the oceans.

Table 4.1: Nutrient Budget of the Oceans

	Nitrogen	*Phosphorus*	*Silicon*
Reserve in ocean	920,000[a]	120,000	4,000,000
Annual use by phytoplankton	9,600	1,300	–
Annual contribution by:			
Rivers (dissolved)	19	2	150
Rivers (suspended)	0	12	4,150
Rain	59	0	0
Annual loss to sediments	9	13	3,800

[a] Units, million metric tonnes.

Source: Martin, 1970.

Nitrogen, phosphorus and silicon occur in sea water in a variety of forms. Concentration and nature of these species in the sea are controlled by physical, biochemical and geochemical processes. The main sources of nutrients to the sea are rock weathering and runoff from land. Atmospheric input through precipitation (rain) also contributes to a major extent in the nitrogen system. Man made activities like dumping of domestic sewage and industrial effluents also play a dominant role of the input of nutrients, particularly in rivers, estuaries and coastal waters.

Among the nutrients, nitrogen and phosphorus are used to form the soft tissue of organisms while silicon is used to build the skeletons of planktonic plants (diatoms) and animals (radiolarians). Nutrients, particularly nitrate (NO_3^-) and phosphate (PO_4^{3-}) are utilised by plankton in surface layer of the oceans and photosynthesise carbohydrates from dissolved carbon dioxide and water, according to the equation:

$$CO_2 + H_2O \longrightarrow (CH_2O) + O_2 \qquad 4.1$$

Light is essential for photosynthesis and hence phytoplankton can grow only in the photic zone which is rarely greater than 200 m depth. It is in this photic zone large quantities of nutrient salts are taken up during the active growth of phytoplankton for the building up of their cellular protoplasm. Their uptake by plants also depends on other factors such as salinity, temperature, etc. Phytoplankton form the base of the oceanic food chains and nutrients move along the chains as grazing and predation take place. They are recycled within the water column by excretion and microbial breakdown of organic particulate matter (detritus). Bacteria acts on this (organic) matter to convert them into simple inorganic form. The death and decay of plants and animal bodies thus lead to the remineralisation of nutrients in the water column. However, those particulates which are more resistant to degradation (either enzymatic or bacterial) slowly sink to the bottom and get incorporated in the sediments. Thus in any water body (sea/ocean) the instantaneous balance of nutrient species represents a dynamic equilibrium between several processes by which their uptake, and regeneration in the water column, and incorporation in the sediments occur. They can be conveniently explained in terms of nutrient cycles.

4.1.0 Nitrogen

Nitrogen is present in several valency states ranging from −3 to +5 in sea water. The principal inorganic forms of nitrogen in the sea, in addition to molecular nitrogen are nitrate (NO_3^-), nitrite (NO_2^-), and ammonium ion (NH_4^+). Small concentrations of other forms of nitrogen such as nitrous oxide (N_2O), hydroxyl amine and urea are also present in sea water. The sea also contains very low concentrations of particulate nitrogen associated with marine organisms and the products of their metabolism and decay.

Rain and land runoff from rivers are the main sources of nitrogen to the oceans. The nitrogen in the atmosphere reacts with oxygen during lightening forming nitric oxide. This undergoes oxidation to nitrate which is dissolved in rain water and reach oceans during precipitation (rain).

Analytical Methods for Determination of Nitrogen Compounds

Among the nitrogen containing species, nitrate, nitrite and ammonia are often determined in sea water using photometric methods. The determination of nitrite (NO_2^-N) in sea water is based on the formation of an intense pink azo dye in acid medium on treatment with sulphanilamide and N-(1-naphthyl) ethylene diamine dihydrochloride. The method is very sensitive and is unaffected by the presence of other constituents normally present in sea water. The absorbance of the resulting pink azo dye solution is measured at 543 nm with a spectrophotometer.

The method for the determination of nitrate (NO_3^--N) in sea water involves its reduction to nitrite and subsequent photometric determination based on the formation of an azo dye (Strickland and Parsons, 1968). Among several reducing agents proposed for this purpose, amalgamated cadmium is found to be the most efficient reductant of nitrate. The method consists of treating an aliquot (100 ml) of sea water with ammonium chloride and passing the solution through an amalgamated cadmium packed in a column. The effluent collected from the column is then treated with sulphanilamide and N-(1-naphthyl) ethylene diamine dihydrochloride and the absorbance of the pink azo dye solution is measured at 543 nm with a spectrophotometer.

Ammonium ion (NH_4^+-N) in sea water is usually determined by indophenol blue method which is most satisfactory because of its sensitivity, reproducibility and freedom from interference by other organic nitrogen compounds. The method essentially consists of formation of monochloroamine by the reaction of ammonium with hypochlorite (or Trione) which on reaction with phenol in alkaline medium in presence of sodium nitroprusside (catalyst) gives an intensely coloured indophenol blue. The absorbance of the solution is measured at 630 nm with a spectrophotometer.

4.1.1 Nitrogen Cycle

The concentration of various nitrogen species particularly NO_3^-, NO_2^- and NH_4^+ in the sea are mainly controlled by biological factors. However, physical effects such as the sinking of dead organisms and upwelling tend to bring a redistribution of these species in the water column. Bacteria, plants and animals play complimentary roles in the various transformations as shown in the cycle (Figure 4.1). Bacteria probably dominate the regenerative processes in which organic nitrogen compounds are converted into inorganic nitrogen species and eventually to nitrate. Phytoplankton normally synthesize their proteins, but bacteria usually use them when organic nitrogen is not available. However, the only contribution of living animals to the nitrogen cycle is the excretion of

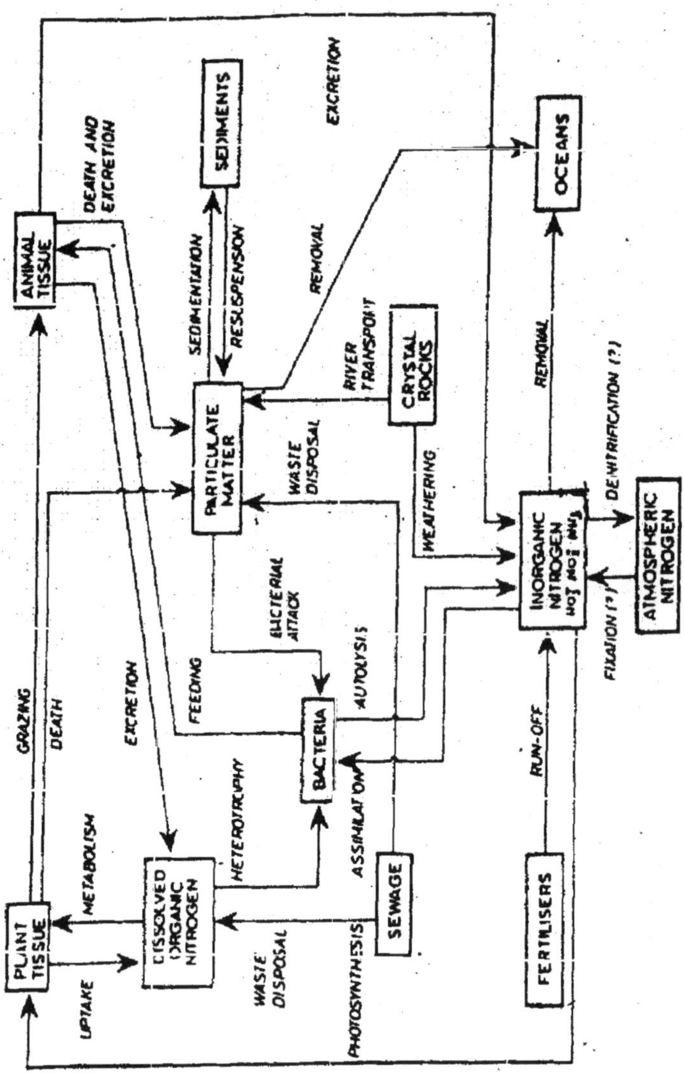

Figure 4.1: The Nitrogen Cycle in the Sea (*Source:* Aston, 1980)

ammonia into the water column and to lesser extent its precursors such as urea, amino acids etc. There are three important stages in the nitrogen cycle. They include (*i*) nitrogen fixation, (*ii*) nitrogen assimilation, and (*iii*) nitrogen regeneration.

(*i*) Nitrogen Fixation

At normal salinity (35.00 psu) and temperature (28°C), the solubility of molecular nitrogen in sea water is about 8.15ml.l^{-1}. In general, sea water is in near saturation with atmospheric nitrogen. Certain blue green algae (*Trichodesmium* spp.) have been shown to fix nitrogen on a large scale in tropical and subtropical waters using solar energy. As a consequence, nitrogen fixation does not occur in darkness. The fixation is also inhibited if an alternative source of inorganic nitrogen (NO_3^-, NO_2^-, NH_4^+) is available which the organisms use in preference to molecular nitrogen.

(*ii*) Nitrogen Assimilation

Phytoplankton utilises NH_4^+, NO_2^- and NO_3^- present in sea water in the order of preference for synthesis of their cellular amino acids. Uptake is confined virtually to the euphotic layer of the sea since it is consequent to photosynthesis. A few cases of phytoplankton *e.g.* phytoflagellates utilise dissolved organic compounds such as urea and amino acids for their nitrogen requirements. Since they are present in sea water at very low concentrations, it is unlikely that they will satisfy the nitrogen requirements of phytoplankton in preference to ammonia and nitrate.

Nitrogen assimilated in the form of nitrate or nitrite by the plankton should first be converted to ammonia before it can be converted into amino acids for protein synthesis. The reduction of NO_3^- to ammonia (which is an endothermic process) proceeds much faster in sunlight in four stages.

$$NO_3^- + 2H^+ + 2e^- \longrightarrow NO_2^- + H_2O \qquad 4.2$$

$$2NO_2^- + 4H^+ + 4e^- \longrightarrow \underset{\text{(hyponitrite)}}{N_2O_2^{2-}} + 2H_2 \qquad 4.3$$

$$N_2O_2^{2-} + 6H^+ + 4e^- \longrightarrow \underset{\text{(hydroxylamine)}}{2NH_2OH} \qquad 4.4$$

$$NH_2OH + 2H^+ + 2e^- \longrightarrow NH_3 + H_2O \qquad 4.5$$

The first step in the reduction (eq 4.2) of NO_3^- is catalysed by a coenzyme II which is produced by the chloroplasts. Ammonia thus produced (eq.4.5) or assimilated directly from sea water is converted into glutamic acid by reaction with ketoglutaric acid in the presence of reduced nicotinamide adenine dinucleotide phosphate (NADPH).

$$NH_3 + HOOC.CO(CH_2)_2COOH + 2NADPH \longrightarrow$$
(ketoglutaric acid)

$$HOOC.CH(NH_2)CH_2.CH_2COOH + 2NADP + H_2O \quad 4.6$$
(glutamic acid)

Other amino acids (numbering about 20) required as building blocks for algal proteins are produced from glutamic acid by transamination. For example, the reaction of glutamic acid with pyruvic acid gives rise to alanine with the reformation of ketoglutaric acid as shown in eq. 4.7:

$$CH_3COCOOH + HOOC\ CH(NH_2)CH_2CH_2COOH \longrightarrow$$
(Pyruvic acid)

$$CH_3CH(NH_2)COOH + HOOC\ CO(CH_2)_2COOH \quad 4.7$$
(Alanine)

Proteins are then produced by linking together of the various amino acids by complicated reaction sequences involving RNA and DNA and the energy provided by adenosine triphosphate (ATP).

(*iii*) Nitrogen Regeneration

The processes by which organic nitrogen compounds are converted into inorganic species are believed to be mainly bacterial and take place in three stages namely (*a*) decomposition of organic nitrogen compounds to yield ammonia, (*b*) nitrification and (*c*) denitrification.

(*a*) Decomposition of Organic Nitrogen Compounds to Yield Ammonia

Soluble nitrogen compounds of dead organisms and those excreted by plants and animals are rapidly broken down to ammonia by various species of proteolytic bacteria which occur at all depths in the sea. Apart from this, the decomposition of particulate organic compounds in the water column also leads to the formation of ammonia. However, some particularly stable organonitrogen compounds resist bacterial attack, and eventually sink to the bottom and are incorporated in the sediment as marine humus.

(*b*) Nitrification

Ammonia thus produced in (a) is subsequently oxidised to nitrite and nitrate. This is termed as nitrification which occurs in the presence of O_2 at all depths, largely mediated by a bacterial mechanism.

$$NH_4^+ + OH^- + 1.5O_2 \rightleftharpoons H^+ + NO_2^- + 2H_2O \quad 4.8$$

$$NO_2^- + 0.5O_2 \rightleftharpoons NO_3^- \quad 4.9$$

(*c*) Denitrification

The process of bacterial reduction of nitrate to NO_2^-, N_2O and molecular nitrogen is called denitrification. Two types of denitrifying bacteria are available in the sea namely (*i*) heterotrophs and (*ii*) autotrophs which utilise organic and inorganic compounds respectively as energy sources for denitrification in the absence of O_2.

$$4NO_3^- + 3CH_4 \longrightarrow 2N_2 + 3CO_2 + 6H_2O \quad 4.10$$

$$(CH_2O) + 4/5\ NO_3^- \longrightarrow 2/5\ N_2 + 4/5\ HCO_3^- + 1/5\ CO_2 + 3/5\ H_2O \quad 4.11$$

$$2NO_3^- + 2H^+ \longrightarrow N_2O + 2O_2 + H_2O \quad 4.12$$

Denitrification is known to occur in the water column as well as in the sediments, provided that the dissolved oxygen concentration is below a critical level (<0.1 ml.l^{-1}) or is absent. This situation may occur at the intermediate depths of the equatorial eastern Pacific and northern Indian Oceans where the dissolved oxygen concentration in the depth range 500-1000 m approaches minimum (almost zero). Denitrification leading to the formation of molecular nitrogen or nitrous oxide (N_2O) in the sea is effective in balancing deficiency in the nitrogen budget.

4.1.2 Distribution of Nitrogen Species in the Sea

(a) Seasonal Variation

Seasonal variation in the concentrations of NO_3^-, NO_2^- and NH_4^+ occurs in surface layers of the sea as a result of biological activity.

These variations are most pronounced in shallow waters near the continents, in mid and high latitudes. The increased intensity and duration of light in spring (Feb/March-April/May) causes a prolification of phytoplankton; leading to rapid removal of dissolved inorganic nitrogen species from the euphotic zone. In summer (May/June-July/Aug) solar heating causes the development of thermocline which inhibits the vertical mixing in the water column. As a consequence, the inorganic nitrogen species rapidly decrease and show minimum during early summer (May-June). The decrease in intensity of uptake of nitrogen by phytoplankton in autumn (Aug/Sept-Oct/Nov) coupled with increase in the regeneration of organic nitrogen into inorganic species in winter (Nov/Dec-Jan/Feb) results in two maxima, the primary maximum during Feb-March and the secondary maximum during Aug-October.

(b) Vertical Variation

Nitrogen compounds (NO_3^-, NO_2^-, NH_4^+) are almost completely utilised by their uptake by phytoplankton and hence show very low or nondetectable levels in the euphotic zone. However, regeneration processes occur throughout the water column below euphotic zone. As a result of this the nitrogen species are recycled within the water column by excretion and microbial break down of organic particulate matter (detritus). Sinking of larger biogenic particles and the vertical movement of zooplankton and other animals feeding on phytoplankton and detritus combine to cause a progressive downward movement of nitrogen species out of the euphotic zone. As a result, nitrate exhibits typical concentration depth profiles showing an increase in its concentration with depth reaching maximum around 1000-2000 m as shown in Figure 4.2b.

4.2.0 Phosphorus

Phosphorus occurs in sea water in a variety of dissolved and particulate forms. In soluble form it mostly exists as orthophosphate. In sea water of average salinity and pH 8.0, about 1 per cent of orthophosphate is present as $H_2PO_4^-$, 87 per cent as HPO_4^{2-} and 12 per cent as PO_4^{3-}. About 96 per cent of the PO_4^{3-} and 44 per cent of the HPO_4^{2-} are apparently present in sea water as ion pairs, probably with calcium and magnesium ions. The dissolved organic phosphorus which constitutes a very small fraction essentially consists of aminophosphoric acids and phosphonucleotides.

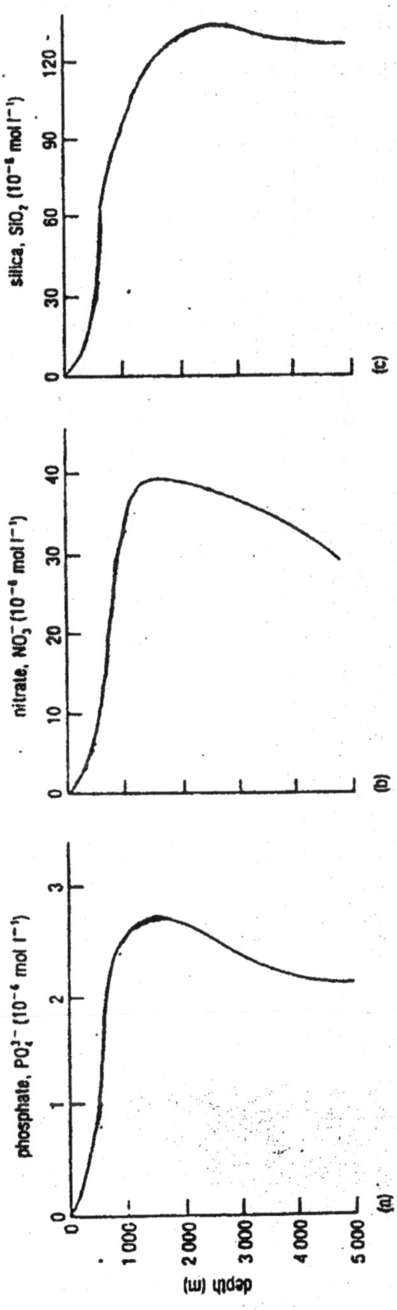

Figure 4.2: Typical Concentration-Depth Profiles for (a) Phosphate, (b) Nitrate and (c) Silicate (*Source*: Open University, 1995)

Particulate matter in the sea contains both inorganic phosphate and organic phosphorus. Some particulate inorganic phosphate is associated with clay minerals and may also occur as solid calcium and ferric phosphates. Phosphate is also likely to be adsorbed on the organic matter and detritus in sea water, and may contain a wide range of phosphorus compounds (in small concentrations) which occur in living tissue.

Determination of Phosphorus in Sea Water

Determination of (reactive) orthophosphate is usually carried out by treatment of an aliquot of sea water sample with acid molybdate reagent containing small amount of antimonyl tartrate and ascorbic acid solution. Absorbance of the resulting blue colour (heteropoly molybdenum blue) is measured at 882 nm using a spectrophotometer. Major cations and anions in sea water and elements such as silicon, arsenic etc. which are capable of forming heteropoly acids with molybdate do not interfere with the reaction under these conditions.

Dissolved organophosphorus compounds do not react with the reagent under these conditions. Hence it is necessary to break down the organophosphorus compounds to (reactive) phosphate. This decomposition can most readily be achieved by treating the sample (sea water) with a small quantity of hydrogen peroxide, and irradiating for few hours with high intensity ultra-violet radiation or by digesting with a strong oxidant, sodium persulphate ($Na_2S_2O_8$). The phosphorus (total) is then determined using the same molybdenum blue method. Organically bound phosphorus is computed from the difference between the total phosphorus and reactive phosphate concentrations.

4.2.1 Phosphorus Cycle

Distribution of various forms of phosphorus in the sea is controlled by biological and physical processes as in the case of nitrogen. Phosphorus cycle is quite simple when compared with nitrogen since the former exists in only one oxidation (+5) state. When dead organisms sink to the sea floor, much of their phosphorus will be subsequently regenerated to the water. However, a portion of it will be incorporated in the sediment as phosphate mineral such as apatite [$Ca_5(PO_4)_3(OH,F)$]. This loss of phosphorus from the sea is balanced by phosphate produced during weathering and transported by rivers as shown earlier in phosphate budget (Table 4.1). The phosphorus cycle essentially consists of two stages namely (i) phosphate uptake, and (ii) regeneration (Figure 4.3).

(*i*) Uptake of Phosphorus

Phosphorus compounds, such as adenosine triphosphate (ATP) and nucleotide coenzymes, play a key role in photosynthesis and other processes in plants. Phytoplankton normally satisfy their requirements of the element by direct assimilation of orthophosphate. Absorption and conversion to organophosphorus compounds proceed even in dark. Bacteria normally satisfy their phosphorus requirements from the organic detritus on which they live. However, if this food source is not rich enough in phosphorus, they are able to assimilate inorganic phosphate from the sea water.

(*ii*) Regeneration of Phosphorus

When phytoplankton and bacteria die, the organic phosphorus in their tissues is rapidly converted to orthophosphate through the agency of phosphatases in their cells. Most of the phytoplankton in the sea are consumed by zooplankton. Part of the assimilated phosphate by zooplankton may return to the sea as excretory products (urine and faecal pellets) which on rapid hydrolysis through the action of phosphorylases are converted into orthophosphate. Any undecomposed organic phosphorus compounds left in the water are rapidly broken down to orthophosphate either by bacteria or by enzymes and return to the water column.

4.2.2 Distribution of Phosphorus in the Sea

(*a*) Seasonal Variation

Seasonal variation in the concentration of orthophosphate in coastal (surface) waters of temperate regions are more pronounced than tropical regions since the latter contain relatively low concentrations of orthophosphate (0.02-0.16 μmol $P.l^{-1}$) at surface. The seasonal variation of phosphate in general, exhibits a similar pattern to that of nitrate. It shows minimum in summer (May-June) during the period of intense phytoplankton productivity and maximum during winter (Dec-Feb) when the regeneration of phosphate dominate over its utilization in photosynthesis. There is

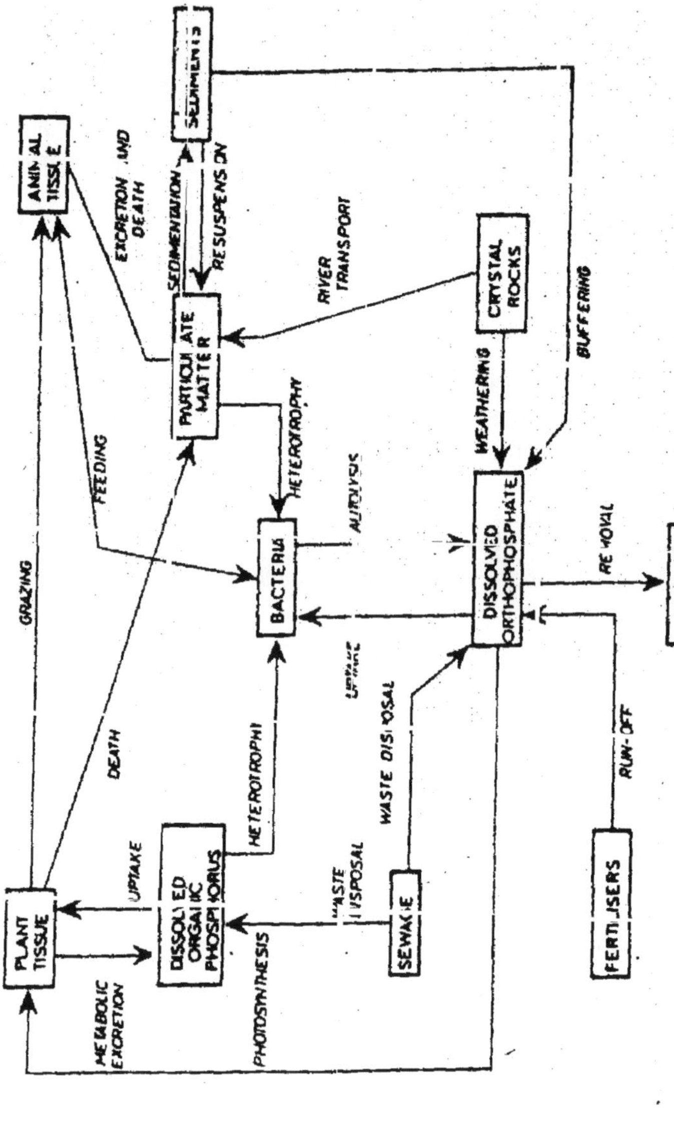

Figure 4.3: The Phosphorus Cycle in the Sea (Sou

an unending argument over whether nitrate or phosphate acts as a limiting factor for phytoplankton production in the sea.

(*b*) Vertical Distribution

Vertical distribution of orthophosphate in the open oceans closely resembles that of nitrate (Figure 4.2a) as it is regulated by the same agencies. The concentration in the surface layer depends on the amount of exchange with the deeper waters. It is relatively high in regions of upwelling where the nutrient rich bottom waters are transported to the surface. The concentration of phosphate increases rapidly with depth reaching to a maximum (500-1000 m) in the neighbourhood of permanent thermocline as a result of oxidation of detritus falling from the surface. Phosphate maximum usually lies close to, or just below the oxygen minimum (Figure 3.3) and the total carbondioxide (CO_2) maximum (Figure 3.9). At higher depths, the phosphate concentration remains more or less uniform upto the bottom. Dissolved organic phosphorus may form a large proportion (upto 50 per cent) of the total phosphorus in surface layer of the oceans. However, this proportion decreases rapidly with depth and is generally insignificant at depths greater than 1000 m. Close to the seabed total phosphorus equals inorganic phosphorus in the open ocean.

Nitrogen and phosphorus are assimilated from sea water in (approximately) constant proportion of 16:1 (by atoms) by phytoplankton as they grow. It is remarkable that ocean waters at all depths usually contain these elements in a similar ratio. This is known as the Redfield ratio (1963). However, there are few exceptions, as the ratio is often low in coastal waters and may show also a seasonal variation. For example, the N:P ratio in the English Channel water varied from 10.5:1 in winter to 19:1 in summer. No such variation exists in warm tropical waters.

4.3.0 Silicon

Siliconis present in seawater both indissolved andparticulate forms. In dissolved form it probably exists as orthosilicic acid [$Si(OH)_4$]. In particulate matter it essentially occurs as weathering materials such as quartz, feldspars, clay minerals and biologically produced hydrated silica (opal).

The silicon budget (Table 4.1) indicates an imbalance of 500 million tonnes between the rate of supply of silicon to the oceans and its rate of removal to the sediments. If the silicon content of the oceans corresponds to a steady concentration, it is necessary to seek other, presumably nonbiological mechanisms for the removal of 500 million tonnes of silicon annually. Reaction of dissolved silicon with degraded aluminosilicates has been postulated as the most probable mechanism of its removal in the oceans.

Determination of Silicon in Sea Water

The method is based on the formation of silicomolybdate complex when an aliquot of sea water is treated with an acid molybdate reagent which on reduction with ascorbic acid gives an intense molybdenum blue complex. Absorbance of the blue colour is measured at 812 nm using a spectrophotometer. Phosphate also gives a similar blue complex, but its interference can be prevented by the addition of oxalic acid.

4.3.1 Silicon Cycle

Concentration of silicon in sea water is affected by both biological and geological processes. The silicon cycle (Figure 4.4) as in the case of phosphorus, essentially consists of two stages namely (*a*) uptake of silicon, and (*b*) regeneration.

(*a*) Uptake of Silicon

Several plants (diatoms and some chlorophyta) and animals (radiolarians, pteropods and sponges) present in the sea have silicified structures. Among them the diatoms are by far the most important organisms in the sea. The exact mechanism by which uptake of silicic acid by the organisms leading to its deposition as hydrated silica on their body parts is not yet known. It is believed that after silicic acid has been adsorbed on a protein monolayer, it polymerises and forms a rigid structure, probably because of hydrogen bonding between the imino groups of the protein and the hydroxyl groups of the silicic acid. Silica thus incorporated in the rigid structure of organisms(such as diatoms) remains insoluble in sea water as long as the organisms are alive, but dissolves fairly rapidly when they die. This may perhaps be due to the thin coating of insoluble silicates of aluminium or iron formed after

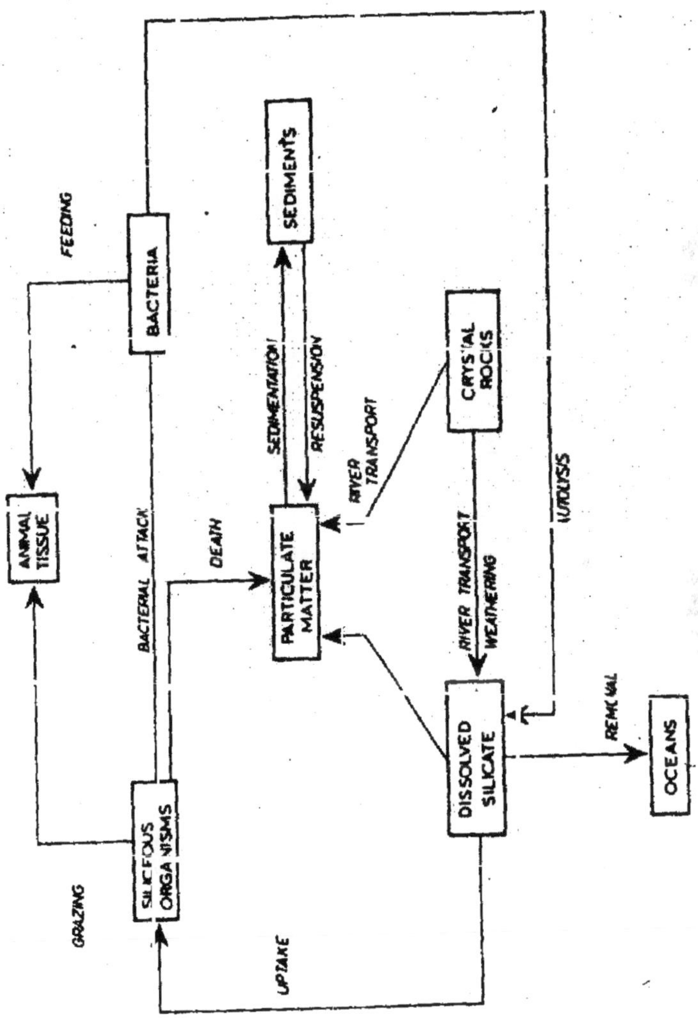

Figure 4.4: The Silicon Cycle in the Sea (*Source:* Aston, 1980)

interaction of these ions or their hydroxides with silica which protects from its dissolution in living siliceous organisms.

(*b*) Regeneration of Silicon

Silicon is dissolved and recycled in the upper parts of the water column. The skeletal material it forms with plants and animals, which are more resistant to solution than soft organic tissue, sink to deeper waters.

4.3.2 Distribution of Silicon in the Sea Water

(*a*) Seasonal Variations

The dissolved silicon content of coastal waters is comparatively higher than nitrate and phosphate because of the effect of runoff from land. In regions of upwelling and blooming of phytoplankton, particularly diatoms, marked seasonal variations are found resembling those of phosphate, though they are greater and erratic. The spring out-bursts of phytoplankton growth causes a rapid decrease in the silicon concentration reaching minimum during April-May, although considerable amounts still remain in the water. Regeneration of silicon commences during summer when phytoplankton growth slackens, and continues until the maximum value is obtained in early winter (December– January).

(*b*) Vertical Distribution

The concentration of dissolved silicon in ocean surface waters is generally low, except in regions of upwelling. In deeper layers, there is a rapid increase in its concentration as in the case of nitrate and phosphate (Figure 4.2c). However, the general distribution pattern differs from one ocean to the other and is determined by the water circulation, and dissolution of diatom skeletons falling from the upper layer.

The depth profiles of important nutrients (nitrate, phosphate and silicon) in the major oceans (Atlantic, Pacific and Indian) are shown in Figure 4.5. Though they exhibit a similar behaviour in all the three oceans, there are subtle differences in the maximum concentration in deeper waters. For example, the maximum concentration of nitrate in deep waters follow the order Indian > Pacific > Atlantic Oceans. On the other hand, maximum concentration of phosphate and silicon follows the order

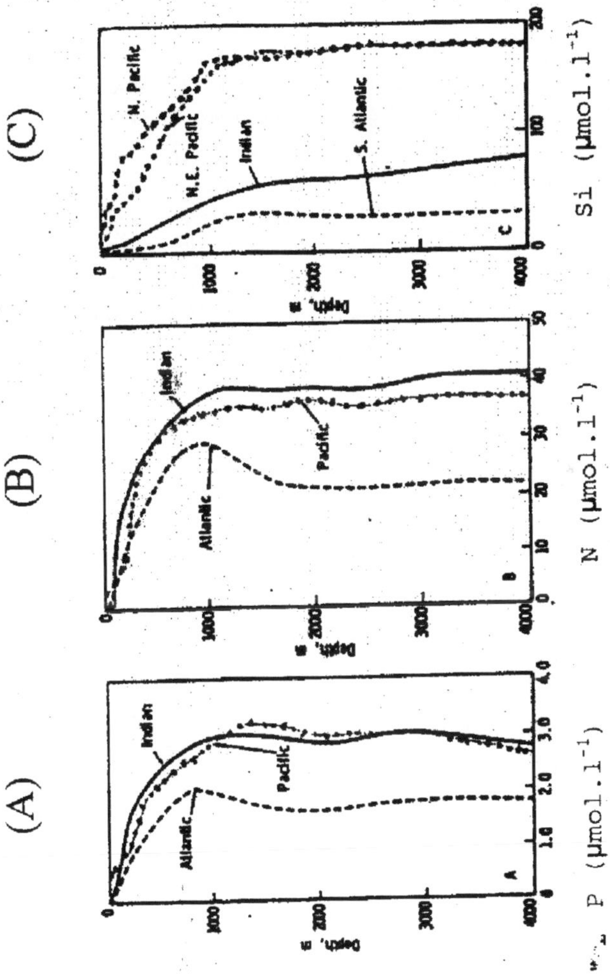

Figure 4.5: Vertical Distribution of Phosphate (A), Nitrate (B)and Silicate (C) in the Atlantic, Pacific and Indian Oceans (*Source:* Horne, 1969)

Pacific>Indian> Atlantic Oceans. This may probably be due to the differences in the exchange of bottom waters (particularly in case of phosphate and silicon) where it is more efficient in Atlantic compared to Pacific and Indian Oceans.

4.4.0 Distribution of Nutrients in the Indian Seas

4.4.1 Seasonal Variation

Concentrations of nutrients (nitrate, phosphate and silicon) in the surface waters of the northern Indian Ocean (Arabian Sea and the Bay of Bengal) exhibit seasonal variations especially in the coastal regions where the upwelling take place. High concentrations of nutrients have been observed along the southwest coast of India (Arabian Sea) and along the east coast of Visakhapatnam during SW monsoon (June-September) season. Concentration levels of nutrients are lowest during the premonsoon (March-May) season which is the period of high phytoplankton productivity and high vertical stability.

4.4.2 Spatial Variation

Distribution of nutrients (PO_4-P and NO_3-N) in the Indian Ocean is widely variable (Figure 4.6). Surface waters in general, are impoverished due to persistent thermocline which inhibits nutrient supply to the euphotic zone from the subsurface layers. However, enrichment of NO_3-N and PO_4-P in surface layers do occur in selected areas such as off Somalia and northwestern Arabian coasts. Which are the strong upwelling zones. By contrast, their concentrations off the east coast of India, where upwelling is either weak or absent, are much lower. Similarly surface waters of the equatorial Indian Ocean contain much lower concentrations unlike their counterparts in the Pacific Ocean because of the absence of wide spread upwelling in the former.

The occurrence of a permanent halocline in the north Indian Ocean together with high primary production in the euphotic zone results in the depletion of nutrients above the thermocline. However, the two-layered circulation leads to their active recycling which results in high organic production at surface. It also leads to low dissolved oxygen and high nutrient concentrations below the thermocline. Profiles of phosphate and nitrate (Figures 4.7 and 4.8) show maxima between 1000-1500 m somewhat below the deep oxygen minimum.

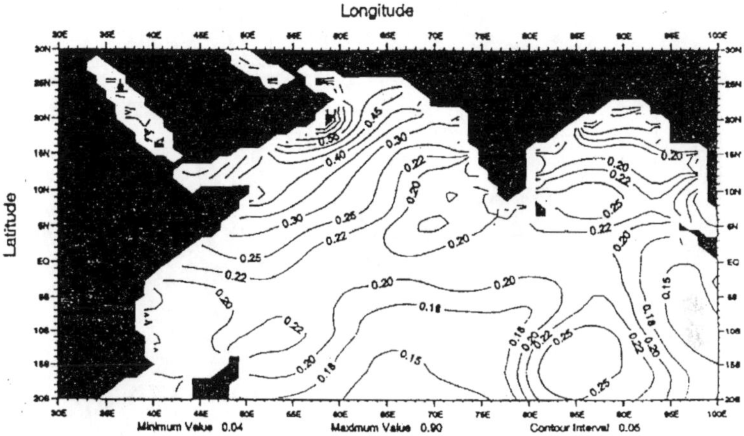

(a)

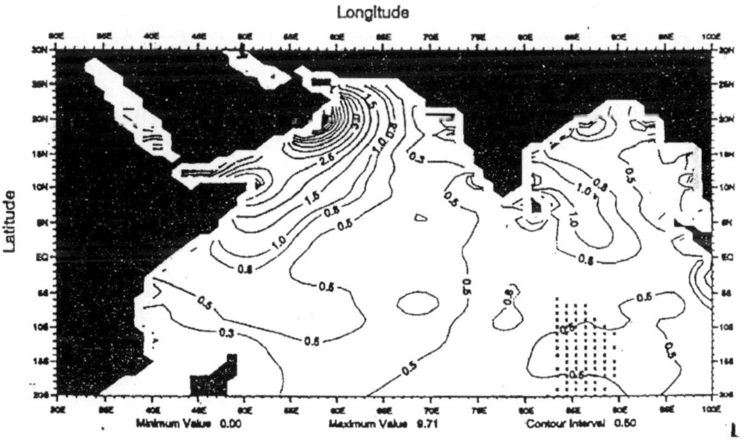

(b)

Figure 4.6: Annual Mean Sea Surface Distribution of Nutrients (mol.l–1) in the North Indian Ocean (a) PO_4–P and (b) NO_3–N (*Source*: Naqvi, 2001)

The concentration of phosphate and silicate are somewhat higher in the Arabian Sea than in the Bay of Bengal. However, nitrate concentrations are slightly lower in the Arabian Sea than in the Bay of Bengal. Silicate in general, shows a steady increase with depth (Figure 4.9) in both the seas.

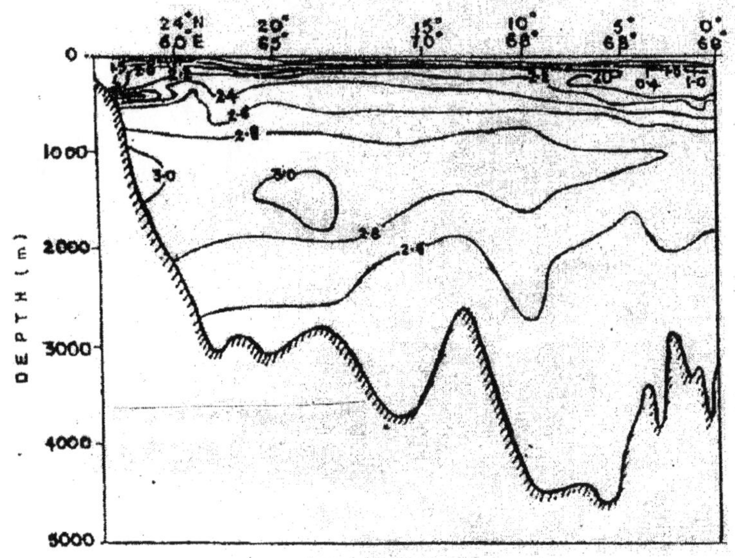

Figure 4.7: Distribution of Inorganic Phosphate (mol.l–1) Along the Section Running from the Gulf of Oman through the Arabian Sea to the Equator
(*Source*: Sen Gupta and Naqvi, 1984)

4.5.0 Vertical and Lateral Movement of Dissolved Nutrients: The Two Box Model

4.5.1 Vertical Movements

Oceans in general, can be treated as layered series of well mixed reservoirs. For convenience, they can be divided into two reservoirs (*i*) a thin upper layer of warm water and (*ii*) a larger bottom reservoir of cold water beneath it. This is necessary to quantify the basic processes occurring in the oceans, and to estimate, to a first approximation, the rate at which different constituents move through the system. The upper box (100-200 m depth) is a well mixed surface layer. Below it lies the lower box (20 times bigger than the upper box) comprising permanent thermocline extending down to 500-1000 m over most of the world oceans followed by intermediate and bottom watermasses. The two box model (Figure 4.10) is further

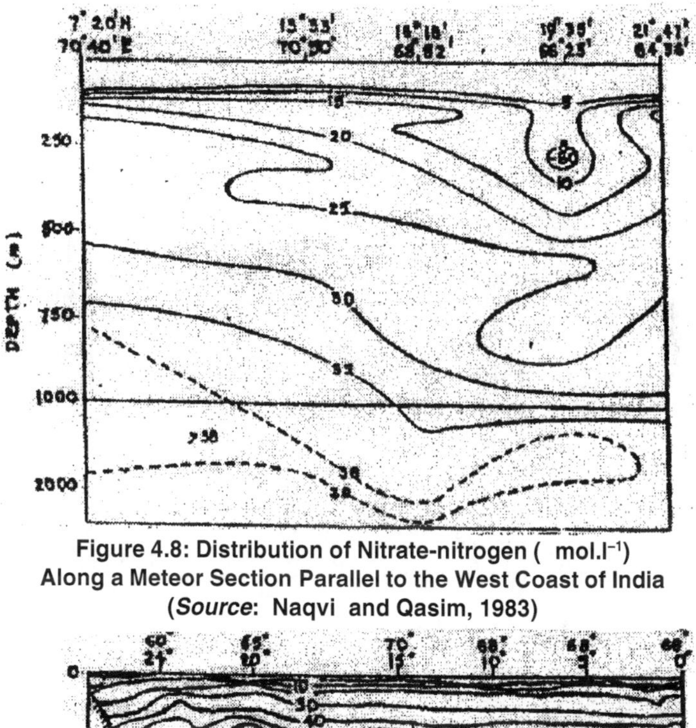

Figure 4.8: Distribution of Nitrate-nitrogen (mol.l^{-1})
Along a Meteor Section Parallel to the West Coast of India
(*Source*: Naqvi and Qasim, 1983)

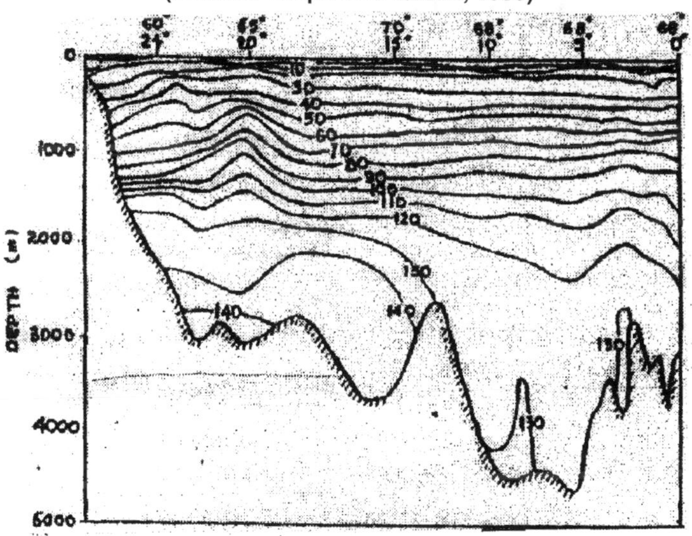

Figure 4.9: Distribution of Reactive Silicate (mol.l^{-1})
Along the Section Running from the Gulf of Oman through
the Arabian Sea to the Equator
(*Source*: Sen Gupta and Naqvi, 1984)

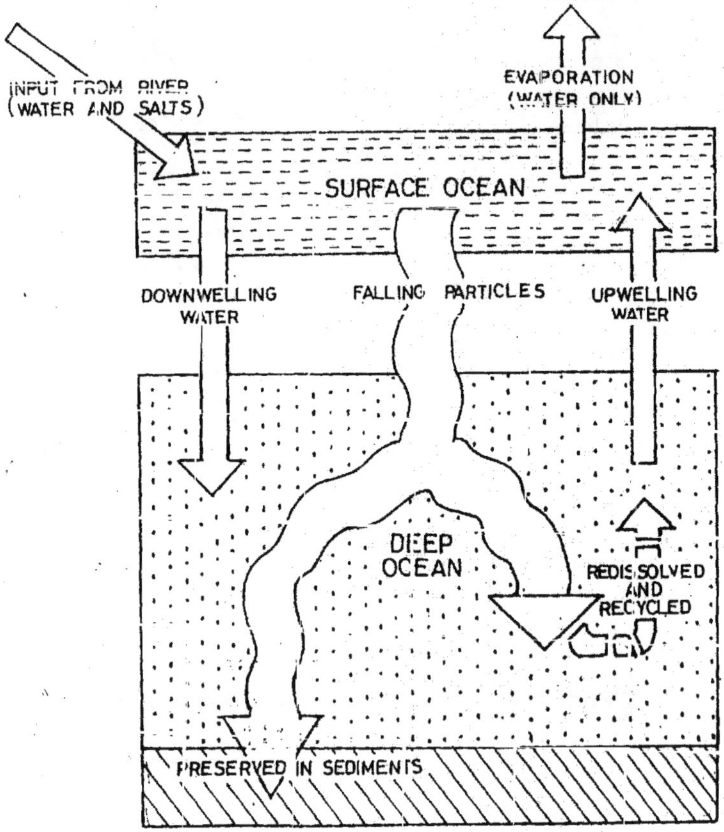

Figure 4.10: Simple two Box Model for the Oceans (*Source*: Open University, 1995a)

simplified by the following assumptions, in order to use it more conveniently.

1. Rivers are the only sources of dissolved constituents to the oceans. Others such as aeolian, volcanic or hydrothermal inputs or the inflow of ground water along continental margins are ignored.
2. The only way dissolved constituents are removed from the oceans is by the organic (biogenic) particles falling to the sea floor.

3. Oceans are in a steady state: The rates of input and losses of any dissolved constituent, both in the ocean as a whole and between the warm and cold reservoirs, have remained constant over long periods so that the concentration at any point do not change with time.

The two box model for phosphate is shown in Figure 4.11. Input of dissolved phosphate is calculated from the data of annual fluxes of water from all the world' s rivers (3.75×10^{16} kg) into the oceans and the average concentration of phosphate in river water (0.5×10^{-6} mol.kg^{-1}). Thus, input from rivers = $3.75 \times 10^{16} \times 0.5 \times 10^{-6} = 18.75 \times 10^{9}$ mol PO_4.

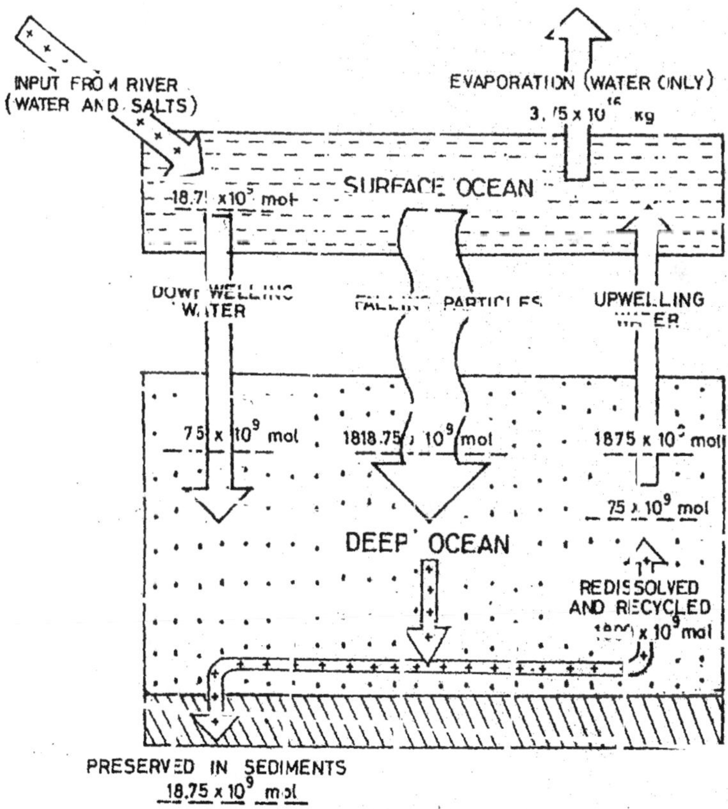

Figure 4.11: Two Box Model for Phosphate (*Source*: Open University, 1995a)

Amount of phosphate preserved in the sediment is 18.75×10^9 mol, since the input from rivers must balance with the amount preserved in the sediment to maintain a steady state composition of phosphate in the oceans.

Based on carbon-14 data, the amount of upwelling flux of sea water is about 20 times the river flux (*i.e.*) $3.75 \times 10^{16} \times 20 = 7.5 \times 10^{17}$ $kg.y^{-1}$. This should be same as that of the downwelling flux. Further, the concentration of phosphate at surface (0.1×10^{-6} $mol.kg^{-1}$) is 25 times less than the concentration at deep ocean (2.5×10^{-6} $mol.kg^{-1}$). Hence, the respective phosphate fluxes can be calculated as follows:

Downwelling water = $7.5 \times 10^{17} \times 0.1 \times 10^{-6} = 75 \times 10^9$ mol PO_4

Upwelling Water = $7.5 \times 10^{17} \times 2.5 \times 10^{-6} = 1875 \times 10^9$ mol PO_4.

If the concentration in the surface ocean is to be maintained at its low level, then the huge difference in the phosphate fluxes for upwelling and downwelling waters must be balanced. Most of the phosphate that is added from rivers and that upwelled from the deep ocean to the surface is fixed by organisms which subsequently die and sink, carrying the phosphate with them. So they transport the balance of the phosphate that is not carried to the deep ocean by downwelling water. The amount of phosphate carried down by falling particles must therefore be given by:

PO_4^{3-} in particles = PO_4^{3-} in upwelling + PO_4^{3-} in rivers – PO_4^{3-} in downwelling.

= $(1875 \times 10^9) + (18.75 \times 10^9) - (75 \times 10^9)$

= 1818.75×10^9 mol PO_4^{-3}.

The amount of phosphate that is redissolved and recycled in the deep ocean is obtained from the difference between the amount of phosphate carried down by the falling particles, and the amount that is preserved in the sediment. Hence:

PO_4^{3-} redissolved and recycled in the deep ocean

= $(1818.75 \times 10^9) - (18.75 \times 10^9)$

= 1800×10^9 mol

This can be checked with the amount of phosphate involved in the upwelling as follows:

$$PO_4^{3-} \text{ in upwelling} = PO_4^{3-} \text{ redissolved and recycled} + PO_4^{3-} \text{ downwelling}$$

$$(1875 \times 10^9) = (1800 \times 10^9) + (75 \times 10^9)$$

In order to make useful comparison of results of two box model calculations, we have to calculate the following:

(i) The Percentage of Phosphorus Entering the Surface Ocean that goes into the Falling Particles

$$= \frac{\text{Amount of falling particles}}{\text{Amount entering the surface (rivers and upwelling)}} \times 100$$

$$= \frac{1818.75 \times 10^9}{(18.75 + 1875) \times 10^9} \times 100$$

= 96 per cent

(ii) The Percentage of Phosphorus Entering the Surface Ocean that Preserved in the Sediment

$$= \frac{\text{Amount preserved in sediment}}{\text{Amount entering surface ocean}} \times 100$$

$$= \frac{18.75 \times 10^9}{1818.75 \times 10^9} \times 100$$

= 1 per cent

(iii) The Percentage of Phosphorus in Falling Particles that Preserved in Sediments

$$= \frac{\text{Amount preserved in sediment}}{\text{Amount in falling particles}} \times 100$$

$$= \frac{18.75 \times 10^9}{1818.75 \times 10^9} \times 100$$

= 1 per cent

The above calculations show that nearly all the phosphate that enters the surface ocean is removed in falling particles. However, only about 1 per cent of the falling particles survives to be preserved in sediments. The rest 99 per cent is redissolved and recycled. Further, only 1 per cent of the phosphate entering the surface ocean (from both rivers and upwelling) is preserved in sediments. Or in other words, we can say that each mole of phosphate goes to an average of about 100 cycles within the oceans before it is incorporated into the sediments on the ocean floor. Most of the nutrient cycling occurs in the mixed surface layer (the upper box), because (photosynthetic productivity) can take place only within the photic zone. Below this, organic matter is consumed (oxidized) and returning the nutrients to solution.

Carbon-14 measurements indicate that it takes on an average about 1600 years for the total volume of water in the oceans to exchange between the deep and the surface reservoirs. If during each one of the exchange cycles, 1 per cent of phosphorus present in the oceans is removed then it would probably take 100 such cycles to remove all the phosphorus. This gives a residence time of phosphorus in the oceans as $100 \times 1600 = 1{,}60{,}000$ years. Residence time of phosphorus can also be calculated from the data on the annual input of phosphorus from rivers (18.75×10^9 mol of PO^4, Figure 4.11) and the total mass of phosphorus (7.93×10^{10} tonnes, Table 1.1). Since each mole of phosphate contain one atom (31 g) of phosphorus, the annual input of phosphorus by rivers is $18.75 \text{x} 10^9 \text{x} 31 = 580 \times 10^9$ g of P which is equal to 580×10^3 tonnes. Hence:

$$\text{Residence time of phosphorus} = \frac{7.93 \times 10^{10}}{580 \times 10^3} = 1{,}40{,}000 \text{ years}$$

Though there is variation in both the estimates because of approximations and assumptions made in the two box model, it is clear that the basic principles are valid.

If appropriate data are available, the two box model can be used to describe the behaviour of any biolimiting element. Results for other nutrient elements, namely, nitrogen and silicon are much the same as those of phosphorus (though the picture of nitrogen is more complicated as indicated in 4.1.1).

4.5.2 Lateral Movements of Nutrients

It is evident from the above discussion that the two box model of vertical fluxes do not pay any attention on lateral variation (advection). On the other hand, it gives an impression that the small proportion of phosphate leaving the cycle each year to be preserved in the sediments, is uniformly spread throughout the oceans. But it is actually not so as is evident from the distribution of phosphate in deep oceans (Figure 4.12). The phosphate concentration varies from a minimum in the north Atlantic and increases southwards, and continues to increase around southern Africa, eastwards and north eastwards in the Indian Ocean, and also eastwards and north eastwards in the Pacific. Table 4.2 summarises concentration ratios in the Atlantic and Pacific Oceans for elements used in biological processes. Since the data gives the ratio (the difference between deep and surface concentration) of each element in Pacific and Atlantic Oceans, it can be considered as a measure of the lateral enrichment of each element in deep Pacific when compared with deep Atlantic waters. It is obvious from the data (Table 4.2) that there is a striking difference in the enrichment of these elements. For example, nitrogen and phosphorus have the lowest ratios (2) because they form soft organic tissue and are mainly recycled in the surface waters. Carbon has an intermediate ratio (3) since it forms both organic tissue and calcium carbonate. The largest ratio (5) for silicon is due to its involvement in skeletal material that mostly dissolves in deep water.

The mechanism of enrichment of nutrients in deep Pacific compared to the cold deep Atlantic Ocean waters is summarised in Figure 4.13(a). Nutrient poor North Atlantic Deep Water (NADW) sinking in the Norwegian sea between Iceland and Norway flows southward in the western Atlantic and is steadily enriched with nutrients derived from the rain of particulate organic matter sinking from the surface and being redissolved in the deep ocean. It then flows round the southern Africa, and into the Indian and Pacific Oceans. In the southern Atlantic it is joined by Antarctic Bottom Water (AABW), which is more nutrient rich than NADW to start with, because of the upwelling along the Antarctic Divergence. Although some AABW flows north in the Atlantic, much of it flows into the Indian and Pacific Oceans, along with the NADW, and nutrient enrichment by sinking particulate organic matter continues.

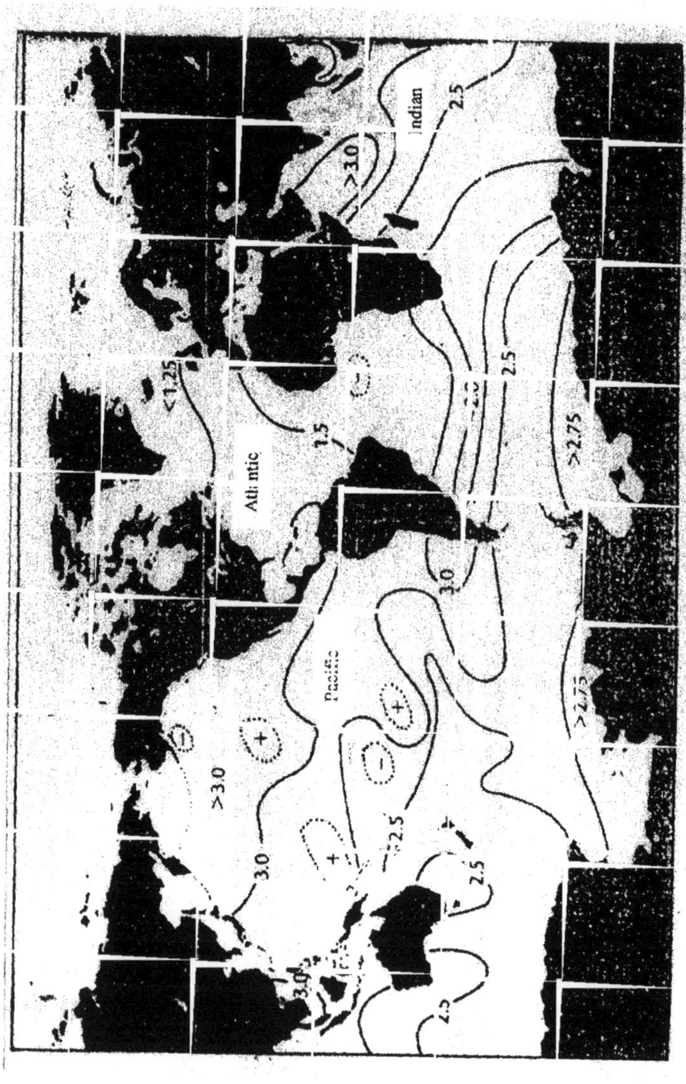

Figure 4.12: Distribution of Phosphate in the Oceans at 2000 m Depth. Contours interval 0.25 × 10^{-6} mol.l^{-1} (*Source:* Broecker, 1974)

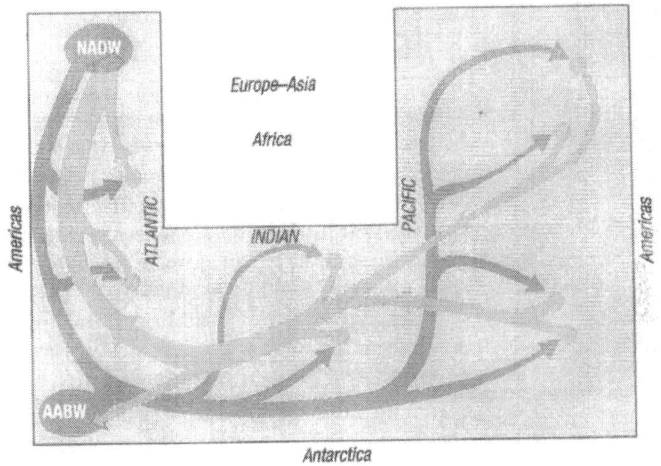

(a)

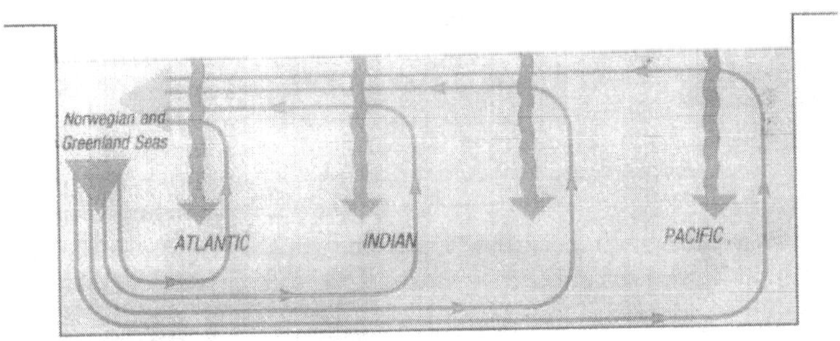

(b)

Figure 4.13: (a) Generalized deep water flow (dark blue) and surface water return (mid-blue) in the oceans. Large ellipses designate sources of North Atlantic Deep Water (NADW) and Antarctic Bottom Water (AABW); small mid-blue circles indicate areas of localized upwelling; (b) Generalized cross-section from the North Atlantic to the North Pacific, showing major advective flow patterns (thin pale blue lines) and the rain of particles (wavy arrows)

(*Source*: Open University, 1995)

Thus, the oldest and the most nutrient rich waters are in the deep Pacific Ocean (Table 4.2). Deep water is returned to the surface mainly by the broad, slow and rather diffuse upwelling that goes on continuously throughout the world oceans. Figure 4.13(b) identifies the localised upwelling areas. However, much of the upwelled water in these regions comes from intermediate depths and is not truly deep water.

Table 4.2: Ratios for Biologically Utilized Elements Between the Atlantic and the Pacific Oceans

Element	$\frac{(C_{deep} - C_{surface})\ Pacific}{(C_{deep} - C_{surface})\ Atlantic}$
Nitrogen (as NO_3)	2
Phosphorus (as PO_{4})	2
Carbon	3
Silicon	5
Barium	7

C: Concentration.

Source: Broecker, 1974.

It should be noted that the simple two box model described for nutrients emphasies only the major processes at work in the oceans. Other aspects such as (*i*) the role of inorganic particles as removal agents, (*ii*) the second order features in the global circulation pattern, and (*iii*) the deviations from the average ratio of one nutrient to another during uptake and release, have not been taken into account. A complete picture of the oceans, must, of course, include all these factors.

Distribution of nutrients and other biologically important elements in sea water must greatly effect the distribution of organically derived sediments. The simple model that we have considered (Figure 4.11), however, assumes only that particulate organic material is either redissolved at depth and eventually returned to the surface, or is preserved for further interaction with sea water by being trapped in sediment. Infact the chemical equilibria that control the behaviour of these elements in deep waters particularly those of calcium and silicon which involve in skeletal body parts, are very important in understanding cycles in deep oceans, as we discussed earlier in carbondioxide-carbonate equilibria (3.3.1).

4.6.0 Stoichiometric Relationships of Nutrients and Oxygen

The concept of apparent oxygen utilisation (AOU) has been introduced by Redfield (1963). In this concept, it is assumed that the oxygen concentration in a watermass which is isolated from the surface will be changed only by biological activity which, in deep water, will be limited to the oxygen consumption involved in remineralisation of organic matter. Assuming that it was originally in equilibrium with the atmosphere, oxygen concentration saturation (O_2') of the water (original) can be calculated from the knowledge of its salinity and temperature. The differenes between the saturation (O_2') and the measured oxygen (O_2) is termed as apparent oxygen utilisation (AOU) and is given by:

$$AOU = O_2' - O_2 \qquad 4.13$$

Any anomaly between the saturated and the measured oxygen concentration (AOU), is ascribed to photosynthesis or biological oxidation of organic matter. Hence, AOU can be related to the relative amounts of nutrient elements and to the carbon in the organic material.

From observation, Redfield (1963) concluded that the ratios of change of carbon, nitrogen and phosphorus in sea water are equivalent to their ratios of concentration in marine phytoplankton.

C: N: P	=	C:N:P
Sea water		Phytoplankton

The question of why nitrogen and phosphorus should occur in sea water in the same ratio that organisms require them remains, as yet, unanswered. We do not know whether the atomic/molar ratio of N : P (16 : 1) occurred at the beginning of evolution and marine organisms adapted to it, or whether the organisms have themselves established the ratio through time.

Redfield established the average carbon: nitrogen: phosphorus molar ratio in phytoplankton as 106 : 16 : 1. A similar molar ratio of change of carbon: nitrogen: phosphorus of 106 : 16 : 1 has also been established for sea water. These correlations lead to a Stoichiometric model for oxidation of organic matter in the ocean represented as:

$$(CH_2O)_{106}(NH_3)_{16}H_3PO_4 + 138\ O_2 \rightleftharpoons 106CO_2 + 16HNO_3 + H_3PO_4 + 122H_2O \qquad 4.14$$

in which the molar formula of organic matter represents the statistical empirical average composition. The significance of this model is that it provides a quantitative basis for the examination of the effects of variations in parameters such as oxygen, nitrogen, phosphorus and to some extent, silicon and carbon dixoide in ocean waters.

The AOU in deep water is related to the amount of inorganic nitrogen and phosphorus released. The total amount of inorganic nitrogen and phosphorus present will include any preformed material in addition to the nutrient salts of oxidation origin (eq. 4.14). Plots of the concentration of nutrients against AOU should be linear and the intercept at zero AOU should correspond to the concentration of preformed nutrients *i.e.* nutrients which existed in the water when they were previously in contact with the atmosphere. Treatment based on these assumptions often agree well with the observed distribution of nutrients and the dissolved oxygen. Such an agreement has been reported for the waters of Atlantic, Pacific and Indian Oceans.

Based on chemical analysis it was calculated (Sen Gupta *et al.*, 1976) that phytoplankton in the Arabian Sea have a molar concentration C:N:P ratio of 105 : 15 : 1. The oxidation of organic matter can, thus, be represented as:

$$(CH_2O)_{105}\ (NH_3)_{15}\ H_3PO_4 + 135\ O_2 \rightleftharpoons 105\ CO_2 + 15\ HNO_3 + H_3PO_4 + 120\ H_2O \qquad 4.15$$

It can be inferred that the ratios of change in sea water may vary because of prevalent chemical process, while the ratios of concentration in marine phytoplankton all over the world oceans will hardly ever vary significantly.

Multiple Choice Questions

1. The nutrient cycling occurs mostly in the sea at
   (a) Surface layer (b) Intermediate layer
   (c) Bottom layer (d) In the sediments ( )

2. Relative enrichment of nutrients, in general, is found in
   (a) Indian Ocean (b) Atlantic Ocean
   (c) Pacific Ocean (d) Antarctic Ocean ( )

3. Denitrification occurs in the sea at
   (a) Surface layer
   (b) Bottom layer
   (c) At oxygen minimum layer
   (d) At salinity maximum layer ( )

4. The ratio of C:N:P in plankton is
   (a) 106:16:1 (b) 200:15:1
   (c) 256:16:1 (d) 106:1:16 ( )

5. The interference of phosphorus in silicon determination using phosphomolybdenum blue method is eliminated by the addition of
   (a) Sulphuric acid (b) Oxalic acid
   (c) Ascorbic acid (d) None of them ( )

6. The major source of nutrients into the oceans is
   (a) Precipitation (rain)
   (b) Land runoff and precipitation
   (c) Atmospheric input
   (d) Anthropogenic input ( )

Short Answer Questions

1. Explain why the vertical profile of phosphate reach maximum at higher depths than nitrate and silicate?

2. Mention the main assumptions made in the two box model for vertical and lateral variation of nutrients.

3. Define apparent oxygen utilisation. How is it related to nutrient concentration in the deep waters?
4. Explain nitrogen budget in the oceans.
5. How do you explain the imbalance in silicon budget of oceans?

Review Questions

1. What are the principal species of nitrogen in the sea? How do you estimate them?
2. Explain the nitrogen cycle in the sea.
3. How do you account for temporal and spatial (vertical) variation of nitrate in the sea?
4. Explain how you determine inorganic phosphate and total dissolved phosphorus in sea water.
5. Describe the salient features of phosphorus cycle.
6. What are the major sources of silicon to the oceans? How do you determine silicon in sea water?
7. Explain the salient features of silicon cycle in the sea.
8. Describe the two box model for vertical and lateral movements of nutrients with special reference to phosphate. Indicate its limitations.
9. How do you account for the interrelationships of nutrients and oxygen in the sea?
10. Compare and contrast the vertical distribution of nutrients (nitrate, phosphate and silicate) in the Atlantic, Pacific and Indian Oceans. How do you account for their enrichment in Pacific when compared to Indian Ocean?

Answers to

Multiple Choice Questions

1. a 2. c 3. c 4. a 5. b 6. b

Chapter 5
Radioactive Nuclides

The atomic weight of any element is the average weight of its isotopes taking into account their abundance. For example, the abundances of the isotopes of potassium, K-39, K-40 and K-41 are 93.10 per cent, 0.012 per cent, and 6.88 per cent respectively. The isotopes of an element may be either stable or unstable (radioactive) which are subjected to decay. Radio isotope (nuclide) of any element has a chemistry identical to that of its stable isotope. Each radionuclide decays, emitting characteristic (, ,) radiation with a characteristic half life.

5.1.0 Types of Radionuclides in the Sea

The radio nuclides present in the sea can be broadly divided into three categories:

1. Long lived primary radionuclides which have existed since the formation of the planet and their short lived daughter nuclides which are continuously renewed by decay.
2. Cosmogenic radionuclides of relatively short half life which are being continuously formed by the interaction of cosmic rays with matter; and

3. Artificial radionuclides produced by human activities, such as fission products from nuclear weapons and waste products from nuclear power plants.

5.1.1 Long Lived Primary Radionuclides

More than 90 per cent of the total radioactivity of sea water arises from ^{40}K (half life 1.30×10^9 years). For sea water of salinity 35 psu which contains 0.40 g K l^{-1}, the radioactivity due to ^{40}K is 331 pci l^{-1} (pico curies l^{-1}). This nuclide decays both by -emission and by K-electron capture, yielding two stable isotopes ^{40}Ar and ^{40}Ca respectively.

$$^{40}_{19}K + {}^{-1}_{0}e \longrightarrow {}^{40}_{18}Ar \tag{5.1}$$

$$^{40}_{19}K \longrightarrow {}^{40}_{20}Ca + {}^{-1}_{0}e \tag{5.2}$$

The decay of ^{40}K to ^{40}Ar provides the basis for one of the most important geochronological methods which will be discussed later in this section.

Natural rubidium (Rb) contains 27.85 per cent of radioactive ^{87}Rb (half life 4.7×10^{10} years). On the basis of rubidium concentration in sea water of 120 μg l^{-1}, the radio activity to sea from this source is 2.9 pci l^{-1} which is less than 1 per cent contributed by ^{40}K.

Much more important from the geochronological standpoint are the members of the Uranium (^{238}U), Thorium (^{232}Th) and Actinouranium (^{235}U) decay series. Each parent nuclide gives rise to a complex chain of decay products as shown in Table 5.1. Natural radionuclides in the series (Table 5.2) show wide range of geochemical behaviour and in general, much lower concentrations than in the earth' s crust. For example, $^{230}Th/^{238}U$ is only 0.01 in the ocean compared to 3 in the rocks. Evidently, there is a complete disruption of the radioactive equilibrium by geochemical processes.

The average concentration of uranium in sea water is 3.3 μg l^{-1} which corresponds to a radioactivity of 2.2 pci l^{-1}. The concentration of uranium in river water widely varies from 0.01-1.2 μg l^{-1}. Its mean global river concentration is estimated as 0.50 μg l^{-1}. The concentration of uranium in deep sea sediments varies over wide range 0.6-6.0 μg g^{-1} with an average of 2-3 μg g^{-1}.

The concentration of thorium in sea water is small (0.05 μg l^{-1}) and in coastal waters it is affected to an appreciable extent, by geochemical processes. Investigation of isotopes of thorium in the marine environment have been mostly concerned with Th-232 and Th-230. Thorium-232 is the parent of the thorium series (Table 5.1). Thorium-230 is the daughter product of uranium-238 with a half life of 7.5×10^4 years.

Protactinium-231 with half life of 3.4×10^4 years (Table 5.2) has close similarity in its behaviour with thorium-230 in sea water. Both the nuclides are rapidly and efficiently removed from sea water on to the sediments. Interest on the comparative behaviour of Pa-231 and Th-230 centers mainly on the use of their ratio in geochronology of sediments.

5.1.2 Cosmogenic Radio Nuclides

Considerable proportion of cosmic ray particles reaching the earth have energies exceeding that binding the nuclei of atoms. Most of their energy is absorbed by the nuclei of the atoms of the atmospheric gases (oxygen, nitrogen and argon) which are thereby fragmented into stable or unstable (radioactive) nuclei, lighter than the parent nuclide. Some nuclides are also produced by neutron capture reactions. All these cosmogenic nuclides are gradually distributed to the lower atmosphere, to the oceans and to the land, the fraction in each of these reservoirs depends on the chemical properties of the particular nuclide. A significant contribution of certain nuclides, *e.g.* ^{26}Al, may occur in extra terrestrial material, such as cosmic dust, which is irradiated in space. The cosmogenic nuclides have half lives ranging from a fraction of a second to more than 10^6 years. So far 20 of them have been detected, but of these only a few are sufficiently long lived to be of value in oceanographic studies (Table 5.3).

Most of the oceanographic work on cosmogenic nuclides has been centred around ^{14}C and ^3H (tritium), both of which are also produced during nuclear explosions. The nuclides are used as indicators for rates of (watermass) mixing processes. Carbon-14 originates through the capture of neutrons by nitrogen in the upper atmosphere as per the reaction:

$$^{14}_{7}N + ^{1}_{0}n \longrightarrow ^{14}_{6}C + ^{1}_{1}H \qquad 5.3$$

The $^{14}_{6}C$ is radioactive and undergoes decay as:

$$^{14}_{6}C \longrightarrow ^{14}_{7}N + ^{-1}_{0}e \qquad 5.4$$

Table 5.1: Important Relationships in the Uranium, Thorium and Actino-uranium Series

Uranium series						
^{238}U (99.274% of U)	451x10^{9} years →	^{234}Th	241 days →	^{234m}Pa	1.17 min →	^{234}U (0.0056% of U)
2.47x10^{5} years →	^{230}Th	7.52x10^{4} years →	^{226}Ra	1602 years →	^{222}Rn	
3.82 days →	^{218}Po	through further short lived intermediates →	^{210}Pb	21 years →	^{210}Bi	
5.01 days →	^{210}Po	138 days →	^{206}Pb (stable)			
Thorium series						
^{232}Th (100% of Th)	1.41x10^{10} years →	^{228}Ra	6.7 years →	^{228}Ac	6.13 hours →	^{228}Th
1.91 years →	^{224}Ra	3.64 days →	^{220}Rn	through further short lived intermediates →	^{208}Pb(stable)	
Actino-uranium series						
^{235}U (0.720% of U)	7.1x10^{8} years →	^{231}Th	25.5 hours →	^{231}Pa	3.25x10^{4} years →	^{227}Ac
21.6 years →	^{227}Th	18.2 days →	^{223}Ra	through further short lived intermediates →	^{207}Pb(stable)	

Source: Burton, 1975.

Table 5.2: Natural Radionuclides Present in the Ocean

Nuclide	*Half Life, Yr*	*Concentration g/ml*	*Isotopic Abundance %*	*Disintegrations per Second per ml*
H^3	1.2×10^1	3.2×10^{-21}	1.0×10^{-16}	1.1×10^{-6}
C^{14}	5.5×10^3	3.1×10^{-17}	1.3×10^{-10}	5.2×10^{-6}
Be^{10}	2.7×10^6	1×10^{-16}	–	7×10^{-8}
K^{40}	1.3×10^8	4.5×10^{-8}	1.2×10^{-2}	1.1×10^{-2} +
Rb^{87}	5.0×10^{10}	3.4×10^{-8}	27.8	1.0×10^{-4}
U^{238}	4.5×10^9	2×10^{-9}	99.3	2.5×10^{-6}
Th^{230}	8.0×10^4	6×10^{-16}	$>3 \times 10^{-3}$	4×10^{-7}
Ra^{226}	1.6×10^3	8×10^{-17}	~ 100	2.9×10^{-6}
U^{235}	7.1×10^8	1.4×10^{-11}	0.7	1.1×10^{-6}
Pa^{231}	3.4×10^4	5×10^{-17}	~ 100	8×10^{-8}
Th^{227} (RdAc)	–	7×10^{-23}	–	8×10^{-8}
Th^{232}	1.4×10^{10}	2×10^{-11}	~ 100	8×10^{-8}
Th^{228} (RdTh)	1.9	4.0×10^{-21}	–	1.2×10^{-7}
Ra^{228} (MsTh)	6.7	1.4×10^{-20}	~ 1×10^{-2}	1.2×10^{-7}

Source: Picciotto,1961.

It is soon converted to carbondioxide and eventually participates in the marine CO_2 cycle discussed earlier (3.3.0). This nuclide is sufficiently long lived (half life 5730 years) for it to be used in the study of water mixing processes. Further, it provides a very useful means of studying the deposition rates of marine sediments.

Tritium (H-3) is produced by interaction of neutrons with nitrogen according to the reaction:

$$^{14}_{7}N + ^{1}_{0}n \longrightarrow ^{12}_{6}C + ^{3}_{1}H \qquad 5.6$$

After formation, most of it is oxidised and falls in rain and snow. The steady state level of the nuclide existing before 1954 in the surface layer of the ocean was 1 atom per 10^{18} atoms of hydrogen. By 1961, tritium produced in thermonuclear weapon tests had raised the average level in the northern hemisphere about 8 folds. However, with the decrease in nuclear testing, the level has subsequently declined somewhat. Because of its relatively short half life (12.3

years), its usefulness is limited to study rapid pocesses such as those occurring in the mixed layer of the oceans.

Table 5.3: Basic Information Concerning Nuclides Produced by Cosmic Rays

Nuclide	^{3}H	^{7}Be	^{10}Be	^{14}C	^{26}Al	^{32}Si
Half–life (years)	12.3	0.145	2.5×10^{6}	5730	7.4×10^{5}	500
Production rate in total atmosphere (atm cm^2s^{-1})	0.25	0.081	0.045	2.5	1.4×10^{-4}	1.6×10^{-4}
Fraction of total earth inventory in:						
Atmosphere	0.072	0.71	3.9×10^{-7}	0.019	1.4×10^{-6}	2.0×10^{-3}
Land surface	0.27	0.08	0.29	0.04	0.29	0.29
Ocean–mixed layer	0.35	0.2	5.7×10^{-6}	0.022	1.4×10^{-5}	0.0035
Ocean–excluding mixed layer	0.3	0.002	10^{-4}	0.92	7×10^{-5}	0.68
Oceanic sediments	0	0	0.71	0.004	0.71	0.028
Average concentration in ocean (10^{-3}dpm kg^{-1} water)	36	–	10^{-3}	260	1.2×10^{-5}	2.4×10^{-2}
Average specific activity in ocean (dpm g^{-1} element)	3.3×10^{-4}	–	1600	10	0.0012	0.008
Global inventory (kg)	3.5	3.2×10^{-3}	4.3×10^{5}	7.5×10^{4}	1.1×10^{3}	1.4
Global inventory (MCi)	35	1.1	6.4	340	0.020	0.023

Source: Burton, 1975.

5.1.3 Artificially Produced Nuclides

The introduction of artificially produced nuclides in the marine environment began in mid 1940′ s with the testing of nuclear weapons in the atmosphere and to a smaller extent, in the oceans. Between 1945 and 1963, weapon tests produced almost 200 megaton equivalents of fission products corresponding to the fission of 2.8×10^{28} atoms of uranium or plutonium. Other main sources of artificial radioactivity is the discharge of low level radio active wastes and activated cooling water from nuclear fuel processing plants and nuclear reactors. Although this has contributed much less overall radioactivity than the fallout from the weapons (Table 5.4), they occur in coastal environments. With the growth of the nuclear power industry, there will be progressive increase in environmental radio activity from this source as shown in Table 5.4. Other minor sources of artificial radio nuclides in the marine environment include accidental release from nuclear plants, dumping of packaged nuclear wastes, discharges from nuclear powered ships (submarines) and encapsulated radio active sources used in technology and medicine. Until now, these have made minor contributions to environmental levels.

Table 5.4: Estimated Total Contribution from Artificial Sources to the Oceanic Inventories of Radioactive Nuclides

Nuclides and Sources	*Estimated Radioactivity (Ci)*	
	in 1970	*in 2000*
Nuclear Explosions (world-wide distribution)		
Fission Products (Exclusive of tritium)	$2\text{-}6 \times 10^8$	9×10^8 *
Tritium	1×10^9	3×10^9 *
Reactors and reprocessing of fuel (Restricted local distribution)		
Fission and activation products (Exclusive of tritium)	3×10^5	3×10^7
Tritium	3×10^5	3×10^8
Total artificial radio-activity	0.4×10^9	1×10^9
Total natural ^{40}K	$\mathbf{5 \times 10^{11}}$	$\mathbf{5 \times 10^{11}}$

* Assumed that atmospheric nuclear testing will continue at about the 1968-70 rate.

Source: FAO, 1971.

More than 200 nuclides of elements ranging in atomic number 30 (Zn) to 66 (Dysprosium) are produced during nuclear explosions. Many of them are shortlived and make no contribution to the long term activity of sea water. Only two nuclides namely ^{90}Sr (half life 28 years) and ^{137}Cs (half life 30 years) have been found to be potentially hazardous in oceans. The radio activity levels of ^{137}Cs were considerably higher in surface waters of the north west Pacific (80-480 × 10^{-12} Ci of $^{137}Cs.l^{-1}$) than in the Atlantic (5–23 × 10^{-12} Ci of $Cs.l^{-1}$) during, and immediately subsequent to the first major period of nuclear testing. The input of ^{90}Sr and ^{137}Cs nuclides at the surface of the ocean, which was essentially free from them earlier, and their penetration into the water column, provides a useful means of studying the rates of mixing of surface and deep waters.

In addition to the fission products, the explosion of nuclear weapons leads to the production of other nuclides due to the capture of high energy neutrons by materials such as bomb casings, earth, water, air etc. where the nuclear explosions take place. The principal of these generally known as the induced nuclides are ^{3}H, ^{14}C, ^{32}P, ^{35}S, ^{51}Cr, ^{54}Mn, ^{55}Fe, ^{59}Fe, ^{58}Co, ^{60}Co and ^{65}Zn. Their relative proportions will vary considerably according to the nature of the materials irradiated. Peak increases of induced nuclides (more than 10-fold and ca 50 per cent respectively) occurred in the northern hemisphere during 1945-1960. However, following cessation of nuclear tests, the amounts have decreased as a result of exchange with natural reservoirs and radio active decay. Tritium and ^{14}C produced by induced activity have considerably augmented the natural levels of the nuclides present in the troposphere.

5.2.0 Use of Radio Nuclides as Tracers

A radioactive isotope of a given element has a chemistry identical to that of the element' s nonradioactive or stable isotope. So it mixes and moves through the ocean just as any other stable isotope does. Because of radioactivity, one can follow the movement of the radio isotope of an element which thus acts as a tracer either in a physical or chemical process. Thus radio isotopes have been extensively used as tracers in the study of (*i*) rates of vertical mixing, (*ii*) rates of gas exchange, (*iii*) geochronology of sediments and (*iv*) growth rates of manganese nodules (Broecker, 1974).

5.2.1 Rates of Vertical Mixing

As pointed earlier, radioactive carbon (C-14) is produced continuously in our atmosphere due to cosmic ray interaction with gaseous nitrogen. About 100 atoms of C-14 are generated over each square centimeter per minute on the earth' s surface. Over many thousands of years, the radioactivity produced by C-14 on the surface of the earth might have reached a steady state since it is disappearing by its own radioactive decay at same rate at which it is produced by cosmic rays. The radioactive decay of C-14 follows the general equation

$$N_t = N_o e^{-} \quad 5.7$$

where,

N_t and N_o are the number of C-14 atoms at time t (in the ocean) and at time t = 0 (when formed originally in the atmosphere) and is a constant characteristic of the C-14 (decay constant). Since C-14 is produced in the atmosphere, it has the highest ratio of C-14/C where C is the ordinary carbon (C-12). On the other hand, the bottom of the deep ocean which is the remote part, has the lowest ratio of C-14/C. Thus the difference in the C-14/C contents between surface which is in contact with atmosphere, and the deep ocean reservoirs (Two box model as discussed in 4.5.0) allows us to estimate the rate of mixing across the main thermocline of the ocean. For this purpose we assume that the surface and deep reservoir sizes and their carbon contents have achieved constant values. Hence between them the distribution of ordinary carbon (C-12) no way depends on the rate of mixing. On the other hand, since C-14 is undergoing decay, its distribution between surface and deep reservoirs depend on the rate of mixing.

Two box model for the cycle of water, carbon (C) and radio carbon (C-14) between surface and deep ocean reservoirs is shown in Figure 5.1. The water conservation equation can be simply written as:

$$V_{up} = V_{down} = V_{mix} \quad 5.8$$

where,

V_{up}, V_{down} and V_{mix} represent water going up, coming down and exchanging between the two reservoirs respectively.

Ordinary carbon (C) is added to the deep reservoir in two ways. It comes down with the descending surface water and it falls from the surface in the form of particles destined for destruction in the deep reservoir. The sum of these two contributions must be exactly balanced by the amount of carbon carried by upwelling water. The balance of these fluxes can be represented as:

$$V_{mix} C_{deep} = V_{mix} C_{surface} + B \quad 5.9$$

where,

C_{deep} and $C_{surface}$ are the carbon contents of upwelling deep water and descending surface water respectively, and B is the carbon added

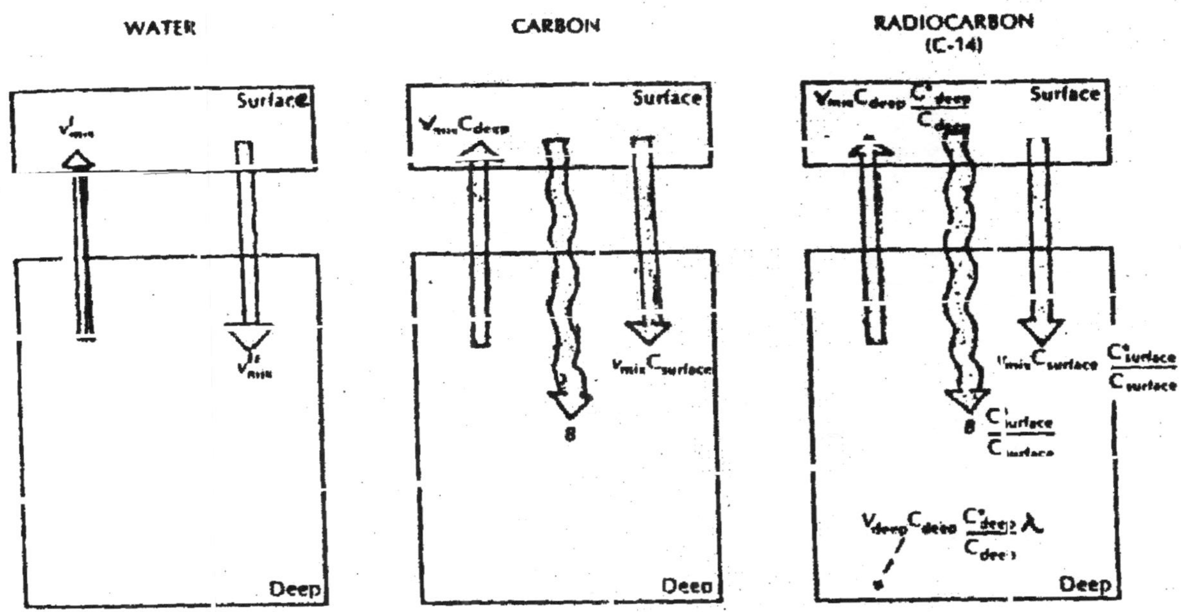

Figure 5.1: The Box Model for the Cycles of Water, Carbon and Radio Carbon (C-14) between the Surface and the Deep Sea. The solid arrows represent the fluxes of substances carried by water and the wavy arrows substances carried by particles. The dashed arrows represents loss by radioactive decay (Source: Broecker, 1974).

to the deep reservoir (each year) by the destruction of particles falling from the surface. Solving for B,

$$B = V_{mix} (C_{deep} - C_{surface}) \qquad 5.10$$

A similar equation can be established for C-14. It is carried up from deep reservoir to surface by upwelling and carried down from the surface to the deep reservoir. In addition the C-14 undergoes decay.

The radio carbon (C-14)in sea water is measured as C-14/C (where C is the normal carbon). In the oceans, for every C-14 atom there are 10^{12} ordinary carbon atoms. Laboratory measurements thus do not give the absolute amount of radio carbon, but the ratio of radio carbon atoms (designated as C^*_{deep} or $C^*_{surface}$) to ordinary carbon atoms (C_{deep} or $C_{surface}$). Hence, the amount of C-14 in descending surface water is given as:

$$V_{mix}\ C_{surface}\ \left[\frac{C^*_{surface}}{C_{surface}}\right] \qquad 5.11$$

The amount of C-14 in upwelling deep water is represented as:

$$V_{mix}\ C_{deep}\ \left[\frac{C^*_{deep}}{C_{deep}}\right] \qquad 5.12$$

Particles carry C-14 down in almost exactly the same proportion to ordinary carbon C, present in surface water. Hence it can be represented as:

$$B\ \frac{C^*_{surface}}{C_{surface}} \qquad 5.13$$

The radioactive decay of C-14, in deep ocean reservoir is represented as:

$$V_{deep}\ C_{deep}\ \left[\frac{C^*_{deep}}{C_{deep}}\right] \qquad 5.14$$

where,

V_{deep} is the volume of water in deep ocean reservoir, and is the decay constant of C-14.

Combining equations 5.11 to 5.14, the conservative equation for C-14 in the deep reservoir can be represented as:

$$V_{mix}\, C_{surface} \frac{C^*_{surface}}{C_{surface}} + B \frac{C^*_{surface}}{C_{surface}} = V_{mix}\, C_{deep} \frac{C^*_{deep}}{C_{deep}} + V_{deep}\, C_{deep} \frac{C^*_{deep}}{C_{deep}} \qquad 5.15$$

Substituting the value of B (eq.5.10) and rearranging the terms, we get:

$$V_{mix}\, C_{deep} \frac{C^*_{surface}}{C_{surface}} = (V_{mix}\, C_{deep} + V_{deep}\, C_{deep}) \frac{C^*_{deep}}{C_{deep}} \qquad 5.16$$

Cancelling the term C_{deep} on both sides and rearranging the eq.5.16, we get:

$$V_{mix} = V_{deep} \frac{\dfrac{C^*_{deep}}{C_{deep}}}{\dfrac{C^*_{surface}\ C^*_{deep}}{C_{surface}\ C_{deep}}} \qquad 5.17$$

The volume of deep ocean reservoir V_{deep} can be expressed in terms of the product of its mean depth (h) and its surface area (A_{ocean}). The equation then becomes:

$$V_{mix} = h\, A_{ocean} \frac{\dfrac{C^*_{deep}}{C_{deep}}}{\dfrac{C^*_{surface}}{C_{surface}}\ \dfrac{C^*_{deep}}{C_{deep}}} \qquad 5.18$$

Dividing both numerator and denominator by C^*_{deep}/C_{deep}, we get:

$$V_{mix} = \frac{h\,A_{ocean}}{\dfrac{C^*_{surface}/C_{surface}}{C^*_{deep}/C_{deep}} - 1} \qquad 5.19$$

Substituting the values of = 1/8200 years, h = 3800 m (3.8×10^5 cm) and

$$\frac{C^*_{surface}/C_{surface}}{C^*_{deep}/C_{deep}} = 1.19 \text{ (for Pacific deep Ocean),}$$

we can find, the annual water exchange (v_{mix}) between the surface and deep ocean reservoirs as:

$$V_{mix} = \frac{\dfrac{1}{8.2 \times 10^3} \times 3.8 \times 10^5}{1.19 - 1}\ A_{ocean} \qquad 5.20$$

$= 240$ cm/year. A_{ocean}

Since the mean thickness of the deep reservoir is 3800 m, transfer of a layer of 240 cm (2.4m) thick between the deep and surface ocean each year yields a mean residence time of water in the deep sea as 3800/2.4 = 1600 years.

5.2.2 The Rates of Gas Exchange

As in the case of vertical mixing, using a simple model, the exchange of gas flux between ocean surface and the atmosphere can be estimated using radio carbon-14. The model assumes that the upper few metres of water column in the sea and the air above the sea surface have uniform concentration of the gas of interest. These two well mixed reservoirs are separated from one another by a stagnant film of water, and the gases move across this boundary layer only by diffusion. Thus if the concentration of the gas in the water is not at equilibrium with the concentration of the gas in the air, there will be a net flow of gas through the stagnant film (as described earlier in section 3.1.1), the rate of which depends on (*i*) the thickness of the film, and (ii) the concentration gradient of the gas between the sea and the atmosphere.

If the partial pressure of the gas above the top of the stagnant film is equal to the concentration of gas below the film then equilibrium would exist and there would be no net transfer of the gas. On the other hand, if the gas concentration on the top of the boundary is higher or lower than the gas concentration in the sea water, then a gradient would be established and the molecules of the gas would be transferred through the stagnant film, the flux (molecules of gas transferred across one square metre in one year) of which is proportional to the concentration gradient. This is expressed by an equation:

$$F = D \frac{[gas]_{top} - [gas]_{bottom}}{Z_{film}} \qquad 5.21$$

where,

F is the gas flux, $[gas]_{top}$ and $[gas]_{bottom}$ are gas concentrations at the top and base of the boundary layer, Z_{film} is the thickness of the boundary layer, and D is the coefficient of molecular diffusion. The diffusion rate depends on the temperature and nature of the gas. It increases with temperature and decreases with the mass of the diffusing molecule. The thickness of the boundary film depends on the wind stress. The greater the turbulance of the sea surface, the more thinner is the film.

The radio C-14 produced in the atmosphere reacts with oxygen forming C-14 O_2 molecules which are transferred into ocean by gas exchange. Since the C-14 has attained a steady state, the number of C-14 atoms added to the oceans every year by gas exchange (as C-14 O_2) with atmosphere must exactly balance the number C-14 that are disappearing within the ocean by radioactive decay. The input of C-14 and loss through its decay are calculated as follows:

$$\text{Input of C-14} = D \frac{[\text{C-14 } O_2]_{top} - [\text{C-14 } O_2]_{bottom}}{Z_{film}} A_{oceans} \qquad 5.22$$

As mentioned earlier C-14 is measured in ratio to normal carbon (C*/C). Hence the C-14 O_2 concentrations are expressed as:

$$[\text{C-14 } O_2]_{bottom} = (C^*/C)_{\text{surface ocean}} [CO_2]_{bottom}, \qquad 5.23$$

and,

$$[\text{C-14 O}_2]_{top} = (C^*/C)_{atm} \frac{\text{C-14 O}_2}{CO_2} [CO_2]_{top} \qquad 5.24$$

where,

$[CO_2]_{bottom}$ and $[CO_2]_{top}$ are the CO_2 gas concentrations at the base and top of the stagnant boundary layer, and the ratio C-14 O_2/ CO_2 accounts for the greater solubility (1.5 per cent) of C-14 O_2 over CO_2. Thus, the C-14/C ratio at the top of the boundary film is slightly greater (1.5 per cent) than the C-14/C ratio measured in atmospheric carbon.

Since there is no net transfer of ordinary carbon between ocean and the atmosphere (because it is assumed to be in the steady state), we can write:

$$[CO_2]_{top} = [CO_2]_{bottom} = [CO_2]_{surface\ ocean} \qquad 5.25$$

Substituting the above values in equation 5.22 and rearranging the terms, we get:

Input of C-14 =

$$\frac{D[CO_2]_{surface\ \ ocean} A_{ocean}}{Z_{film}} \left[\frac{\text{C-14 O}_2}{CO_2} (C^*/C)_{atm} - (C^*/C)_{surface\ \ ocean}\right] \qquad 5.26$$

$$\text{Loss of C-14} = A_{ocean}\, h\, [\quad CO_2]_{ocean}\, (C^*/C)_{ocean} \qquad 5.27$$

where h is the average depth of the ocean and   is the decay constant of radio carbon-14.

Equating the input and decay and solving for Z, we obtain:

$$Z_{film} = \frac{D\,[CO_2]_{surface}}{h\,[\ CO_2]_{ocean}} \; \frac{\left[1 - \dfrac{(C^*/C)_{surface}}{(C^*/C)_{atm}} \quad \dfrac{\text{C-14 O}_2}{CO_2}\right]}{\left[\dfrac{(C^*/C)_{ocean}}{(C^*/C)_{atm}} \quad \dfrac{\text{C-14 O}_2}{CO_2}\right]} \qquad 5.28$$

The average value of diffusion coefficient of CO_2 (D) = $3X10^{-2}$ m^2y^{-1}, the mean depth of the ocean (h) = 3800 m, the reciprocal of the C-14 decay constant (1/) = 8200 y, the average concentration of CO_2 gas in the surface ocean, $[CO_2]_{surface}$ = 0.01 mol.m^3, CO_2 = 2.4 mol.m^{-3}, and the ratio of aC-14O_2/ CO_2 = 1.015. The (C*/C) ratio in surface ocean water is, on the average, 3.5 per cent lower than the (C*/C) ratio in the atmosphere. Substituting these values in equation 5.28, the thickness of the film (Z_{film}) can be obtained as 1.7×10^{-5} m (*i.e.*) 1.7×10^{-3} cm or 17 microns (1 micron = 10^{-4} cm).

Since the molecular diffusion coefficients (D) do not vary greatly from one gas to another, it is convenient to represent gas exchange rates in terms of D/Z. Thus the CO_2 exchange rate (velocity) is represented as:

$$D/Z = \frac{3 \times 10^{-2}}{1.7 \times 10^{-5}} = 1700 \text{ m/y} \qquad 5.29$$

Thus each day a column of sea water of 4-5 m (1700/365) thick exchanges its gas with the atmosphere. In another words on average a gas molecule resides in the surface layer (upper 100 m) of the ocean about 20-25 days before returning to the atmosphere.

5.2.3 The Geochronology of Sediments

Geochronological methods for the determination of rates of sedimentation based on the use of radioactive nuclides are of two kinds. In the first, a time scale is provided for the extent of decay of an unsupported nuclide. If the number of atoms of an unsupported nuclide incorporated in unit mass of sediment at the time of formation is N_o, then the number of atoms remaining at time, t after deposition of the sediment is given by the general equation (eq.5.7).

$N_t = N_o e^{- t}$ where, is the decay constant of the nuclide concerned.

The second kind of dating procedures, utilise the growth of a daughter nuclide in material which initially contain the parent nuclide segregated from negligible amount of the daughter. If the number of atoms of parent and daughter nuclides in unit mass of material after time, t are N and D respectively, and if there is initially a negligible amount of the daughter present, then the time lapsed since the segregation of parent and daughter is given by:

$$t = \frac{1}{\lambda_p - \lambda_d} \ln \left[1 + \frac{D}{N} \frac{(\lambda_p - \lambda_d)}{\lambda_p}\right] \quad 5.30$$

where,

λ_p and λ_d are the decay constants of the parent and the daughter respectively.

Estimation of sedimentation rates in the oceans are based on three primary methods namely (a) use of Carbon-14, (b) use of Thorium-230 and Protactinum-231 which are produced by decay of Uranium dissolved within the sea, and (c) use of Potassium-40 which decays to a gas, Argon-40. These three methods which are used to date deep sea sediments give nearly the same average accumulation rates. Let us describe briefly each method.

We have already discussed the way in which C-14 is generated and mixed with the ordinary carbon (C-12), in the atmosphere-ocean system. Foraminiferan shells and coccolith tests when formed in surface ocean water contain C-14 as well as ordinary carbon atoms in the ratio $1{:}10^{12}$. Once formed, the shells constitute an isolated system and their C-14 content decreases with time as described in eq.5.7. The fundamental assumption in C-14 dating is that the C-14/C ratio in sea water has always been the same as it is today. Hence its decay in the shell can be given by an equation:

$$(C\text{-}14/C)_{today} = (C\text{-}14/C)_{formation} \times e^{-\lambda t} \quad 5.31$$

or

$$t = \frac{1}{\lambda} \ln \frac{(C\text{-}14/C)_{formation}}{(C\text{-}14/C)_{today}} \quad 5.32$$

where,

t is the age of the sample, λ is the decay constant of C-14 and $(C\text{-}14/C)_{formation}$ is the ratio in the present surface ocean water and hence living shells and coccoliths.

The second dating method involves the element uranium and its two radioactive daughter isotopes Th-230 and Pa-231. Th-230 was produced by the decay of long lived U-238 (half life 7 billion years). This was produced from hydrogen in the interior of the stars that subsequently exploded and dispersed their products within our galaxy. It was incorporated in the solar system when it was formed from a galactic dust and gas cloud billions of years ago. Th-230 produced from U-238 has a half life of 75,000 years and it has much different chemical cycle than the latter. While U-238 is quite soluble in sea water, Th-230 is extremely reactive and is removed from sea water on the particulate matter which in turn is incorporated in the sediments. Thus the separation of parent U-238 and daughter Th-230 is the basis for uranium dating. For successful dating by Th-230 method, the rate at which the sediment particles fall onto a given area of the sea floor and the rate at which Th-230 particles are incorporated into the falling sediment particles must remain constant with time.

As in the case of C-14, the amount of Th-230/U-238 in a given weight of the sediment core at a given depth (at time t) can be related to the ratio of Th-230/U-238 in the sediment when it was deposited (at time 0). Thus:

$$(\text{Th-230/U-238})_{\text{present}} = (\text{Th-230/U-238})_{\text{depostion}} \times e^{-\ t} \qquad 5.33$$

where,

is the decay constant of Th-230 and t is the age of the sediment. The reliability of the Th-230 method can be demonstrated from Figure 5.2a which is a plot of Th-230 on a core from the Caribbean Sea. The best fit line corresponds to a sedimentation rate (slope of the line) of 2.35 cm/10^3 years for the core.

The other isotopic product Pa-231 was produced by the decay of U-235 present in sea water. Like U-238, U-235 has a very long half life (0.7 billion years). The U-235 on decay gives Pa-231 with a half life of 34,000 years. Like Th-230, Pa-231 is very insoluble and is removed from sea water to sediments soon after it is produced. Hence a logarithmic plot of the concentration of Pa-231 at various depth intervals in a deep sea sediment core gives a straight line and the slope of the line gives the sedimentation rate of the core as in the case of Th-230. This is illustrated in Figure 5.2b for the same Caribbean core as discussed above. It is evident from the figure that the best fit line for Pa-231 has a slope of 2.35 cm/10^3 years same as that obtained from the Th-230 isotope method. In general, Th-230 allows a mean rate to be established over an interval of 400,000 years while Pa-231, over an interval of 150,000 years.

The third method, K-40–Ar-40 method is used to date deep sea sediments of more than 400,000 years of age. K-40 is generated long

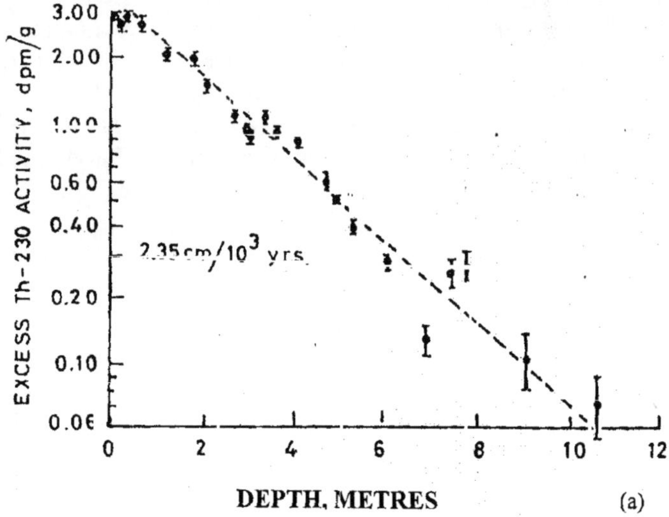

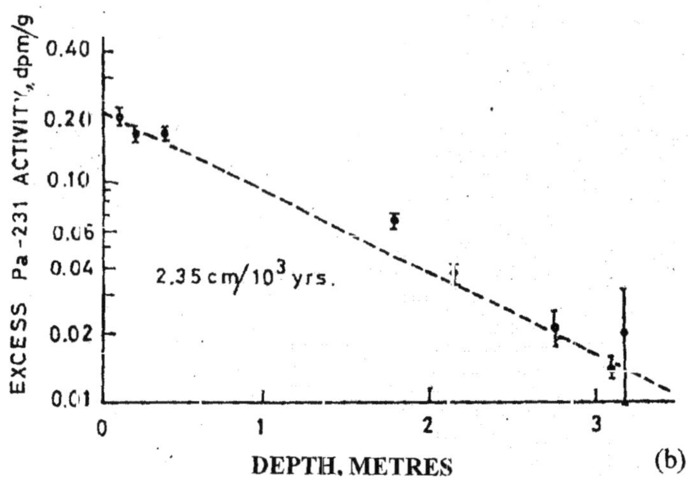

Figure 5.2 (a) Actual Th-230 and (b) Pa-231 Measurements for a Core in Caribbean Sea (*Source*: Broecker, 1974)

ago in the stars and incorporated in our solar system with a half life of over a billion years. Roughly 90 per cent K-40 atoms decay to Ca-40 and the remaining 10 per cent decay to Ar-40. In age determinations, it is this 10 per cent of Ar-40 produced is measured rather than 90 per cent of Ca-40. This is due to the fact that when volcanic rock first forms, gases are almost totally excluded from minerals that crystallize from hot liquids. Hence volcanic rocks are initially free from Ar-40. With time the K-40 in these rocks undergo

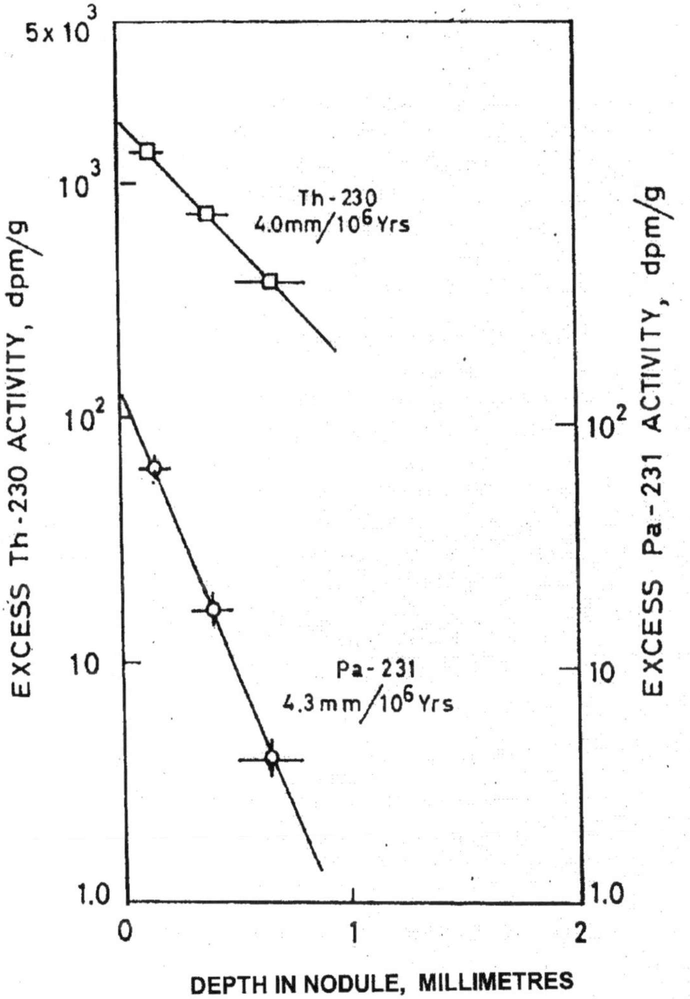

Figure 5.3: Logarithemic Plants of Excess Th-230 and Pa-230 in a Manganese Nodule Dredged from North Pacific (36°N, 160°W; Depth 5400m) (*Source*: Broecker, 1974)

radioactive decay and produce Ar-40 which remains entrapped in the crystal lattice of the rock. If we measure the amount of Ar-40 and the amount of K-40 in one of the rocks, it is possible to calculate the length of time (age) that has elapsed since the rock crystallized.

Application of this method directly to deep sea sediments requires the presence of a layer of volcanic ash. Unfortunately, volcanic ashes in deep sea sediments are quite rare and hence the K-Ar method does not prove to be of much advantage in dating sediments.

5.2.4 Growth Rates of Manganese Nodules

In areas where manganese nodules are found in deep Pacific, Atlantic and Indian Oceans, the accumulation rate of the surrounding sediment averages about 0.3 cm/10^3 years or 3 m/10^6 years. On the other hand, the growth rate of manganese nodules is about 3 mm/10^6 years which is 1000 times slower than the rate of sediment accumulation. It is evident from the above that a nodule with a radius of about 3 cms would have grown over a period of 10 million years.

Several radio isotopes discussed earlier have been employed for the determination of growth rates of manganese nodules. For example, Th-230 and Pa-231 methods gave accumulation rates of about 4 mm/10^6 years (Figure 5.3). Other methods such as U-234 and K-Ar and Be-10 gave accumulation rates of 2-3 mm/10^6 years. Thus all these radio dating methods gave similar accumulation rates for manganese nodules of about 2-4 mm/10^6 years, confirming that the nodules grow at a much slower rate than the accumulation of sediment surrounding them.

Multiple Choice Questions

1. About 90 per cent of radioactivity in the sea is due to

   (a) Rb-87 (b) U-238

   (c) Th-232 (d) K-40 ( )

2. The half life of Carbon-14 is

   (a) 30 years (b) 5700 years

   (c) 12.7 years (d) 28 years ( )

3. The ratio of tritium (H-3) to hydrogen in the atmosphere before II world war was
   (a) $1:10^{12}$ atoms (b) $1:10^{14}$ atoms
   (c) $1:10^{8}$ atoms (d) $1:10^{18}$ atoms ( )

4. The rate of accumulation based on radio dating of manganese nodules is in the order of
   (a) 2-4 mm/10^6 years (b) 5 mm/10^6 years
   (c) 3 cm/10^6 years (d) 3 mm/10^3 years ( )

5. Argon-40 is produced in the volcanic rocks by the decay of
   (a) Uranium-235 (b) Thorium-230
   (c) Potassium-40 (d) Beryllium-10 ( )

6. The half life of Protactinium-231 is
   (a) 7×10^9 years (b) 75,000 years
   (c) 7×10^8 years (d) 34,000 years ( )

Short Answer Questions

1. Name three important radioactive series and indicate their decay sequence.
2. What are cosmogenic nuclides? Name two of them and indicate their significance.
3. How artificial radioactive nuclides are produced? Indicate their importance with few typical examples.
4. Explain the principle and application of geochronology of Carbon-14.
5. Explain how the accumulation rates of manganese nodules can be determined using radio nuclides.
6. Indicate the geochronological applications of Th-230 and Pa-231.
7. Explain the principle involved in K-Ar dating of deep sea sediments. What are its limitations over other dating methods?

8. Explain the principle involved in the determination of CO_2 exchange between sea surface and atmosphere using Carbon-14.

9. Using Carbon-14, explain how you will determine the vertical mixing of water between surface and deep ocean?

Review Questions

1. How do you classify radioactive nucides in the sea? Describe their origin and importance.

2. What is meant by geochronology? Describe the principle and application of radio nuclides used in the determination of sedimentation rates.

3. Explain the principle involved in the use of radio Carbon-14 for the study of rates of vertical mixing and gas exchange in the sea.

Answers to

Multiple Choice Questions

1. d 2. b 3. d 4. a 5. c 6. d

Chapter 6
Organic Matter

Organic matter in the sea can be broadly divided into two categories; dissolved (DOM) and particulate (POM). The former includes true dissolved matter together with colloidal matter passing through a 0.45 μm (1μm=10^{-6} m) filter whereas the latter comprises material having a diameter greater than 0.45 μm and retained on the filter. Thus a widely used procedure for satisfactory separation of dissolved from particulate fraction is filtration through a filter having pores of diameter approximately 0.5 μm. An important distinction between particulate and dissolved organic matter is that the latter cannot sink into the deep ocean, but must be advected or mixed downwards. The amount of DOM in the sea usually exceeds over POM by a factor of about 10-20.

6.1.0 Dissolved Organic Matter (DOM)

The significance of DOM in recent years centres around marine microbiology which is important in the study of phyto and zooplankton. It is now recognized that dissolved organic compounds may influence the state of inorganic substances in sea water and also affect sedimentary processes in the oceans. The DOM which is commonly measured as DOC (dissolved organic carbon) in the oceans, usually lies in the range 0.5-2 mg C l^{-1}, although values as high as 20 mg C l^{-1} may be found in coastal waters, as a result of increased phytoplankton activity and pollution from land. DOC in sea water can be determined by oxidation with peroxy disulphate ($K_2S_2O_8$) in an autoclave or by exposure to intense UV radiation in presence of hydrogen peroxide and measuring the liberated CO_2 either by infrared (IR) absorption or by adsorption chromatography. Similarly dissolved organic nitrogen (DON), dissolved organic phosphorus (DOP) are determined by oxidation of the sample followed by the spectrophotometric determination as combined NO_2+NO_3 for nitrogen and reactive phosphate (PO_4--P) for phosphorus as described in Chapter 4.

6.1.1 Sources of DOM

There are two important sources of organic matter in the sea (*i*) external (allochthonous) and (*ii*) internal (autochthonous). The external sources comprise atmosphere, rivers and sediments while the internal sources consists of living marine organisms including plankton (phyto and zoo) and their extra cellular metabolites, excretions and their dead and decay products. Table 6.1 incorporates data on inputs, reservoirs and losses of DOM in the sea.

The DOM brought into the oceans by precipitation amounts to 2.2×10^{14} g.y^{-1}, an amount comparable with the input of rivers (1.8×10^{14} g.y^{-1}). The major route of chlorine containing pesticides (DDT and its derivates) into the sea is via the atmosphere because of their high volatility. Polychlorinated biphenyls (PCB' s) are also widely found in the marine environment. Since they are less volatile and more water soluble than DDT, they enter the sea from land run off via rivers. DOM in river waters contain humic material (Gelbstoff) which imparts yellow colour to sea water and other decomposition products of vegetable matter from the soil. Soluble organic materials in the form of sewage and industrial effluents also find their entry into the coastal waters through land runoff.

Phytoplankton are the main internal source of DOM in the oceans. It has been estimated that benthic algae accounts for 1/10 th of total production in the sea. The amount of bacterial production in the sea by autotrophs such as photosynthetic and nitrifying bacteria may be regarded as quite insignificant. Organic matter is produced in the sea by chemical or biological processes. There are several routes whereby plankton products enter the DOM fraction. Algae release some of their photosynthetic products from the cell. Zooplankton release DOM and POM as excretion products. Soluble components of algae may leak into the sea during grazing because of cell damage. After death of organism, the autolytic and decay processes will result in the production of soluble compounds. Zooplankton may be responsible for the production of DOM in other ways. Firstly they release soluble organic excretions from their body as a result of internal mechanism; secondly some soluble compounds will be released along with the faeces; finally, the digestive enzymes as well as bacteria from the gut will act on the faecal material to produce DOM in the sea.

Table 6.1: Inputs, Reservoirs and Losses of Organic Material in the Sea

Reservoirs	
Dissolved organic carbon(assuming 700 $g.C.l^{-1}$)	$1x10^{18}$ g.C
Particulate organic carbon(assuming 20 $g.C.l^{-1}$)	$3x10^{16}$ g.C
Plankton	$5x10^{14}$ g.C
Annual inputs	
Net Primary productivity	$3.6x10^{16}$ g.C
(assuming 100 g.C. fixed $m^{-2}yr^{-1}$)	
Rain(assuming 1 $mg.C.l^{-1}$)	$2.2x10^{14}$ g.C
Rivers(assuming 5 $mg.C.l^{-1}$)	$1.8x10^{14}$ g.C
Possible annual inputs into the dissolved fraction	
Phytoplankton excretion (10 per cent of production)	$3.6x10^{15}$ g.C
Resistant material from phytoplankton	$1.8x10^{16}$ g.C
(5 per cent of production)	
Annual losses by sedimentation	
Nearshore sedimentation	$2.7x10^{12}$ g.C
Pelagic sedimentation	$9.2x10^{13}$ g.C

Source: Williams, 1971 and Skopintsev, 1971.

6.1.2 Nature of DOM

Carbon is the major component of organic matter followed by nitrogen and phosphorus. These elements may provide gross organic composition of DOM or POM in sea water. Concentration of carbon (DOC) generally is higher (0.6-2.0 mg C l^{-1}) in the upper 100 m of water column in the oceans. The values for deep water may be about two-thirds (2/3) of the surface waters. Typical concentrations of DON are in the range 5-8 µmol. N l^{-1} in upper 100 m, falling to 2.5 µmol N l^{-1} in the deep ocean. Concentration of DOP is generally very low and fall in the range 0.05-2.0 µmol P l^{-1}. However, the gross elemental composition gives limited information, and for a variety of reasons, information on individual compounds or class of compounds of DOM may be required. These include lipids, hydrocarbons and their halogenated derivates, carbohydrates, amino acids, polypeptides, vitamins, miscellaneous compounds (urea, acetone, butyraldehyde, 2-butanone, nucleic acids) and uncharacterised material, gelbstoff which is a polymeric material containing mixture of high molecular weight compounds humic and fulvic acids..

Lipids are heterogeneous group of biological compounds soluble in nonpolar and semipolar solvents. Their concentration in the Gulf of Mexico varied in the range 0.15-0.31 mg l^{-1}, equivalent to 13-21 per cent of the total DOM. Concentration of lipids in the Arabian Sea off Bombay coast varied in the range 60-140 µg. l^{-1} in dissolved and 20-100 µg l^{-1} in particulate fractions (Nanda Kumar *et al.*, 1987). Both sugars and amino acids are probably important in the turnover of organic material by heterotrophic micro organisms, and in supplying them with energy. Typical concentrations of total carbohydrates in the Indian and Pacific Oceans found to vary from 200-600 and 990-1150 µg l^{-1} respectively. Total dissolved carbohydrate (TCHO) concentrations varied from 72-1150 while particulate carbohydrate (PCHO) varied from 22-125 µg l^{-1} in the central Arabian Sea. However, individual sugar concentrations normally fall in the range of non-detectable to 20 µg l^{-1}. The concentration of individual free amino acids usually fall within the range <1-10 µg l^{-1}. Low molecular weight amino acids (glycine, serine and alanine) are generally dominant among the free amino acids in sea water. Vitamins are the growth regulating organic compounds in the sea. Thiamine is generally required by algae in relatively large amounts than Vitamin B_{12} and accordingly its concentration (8-15 ng l^{-1}) is higher than Vitamin B_{12} (0.1-4.0 ng l^{-1}). Vitamins are generally found in higher concentrations in coastal than in offshore waters. Marine organisms produce extra cellular metabolites which are capable of retarding the growth of other organisms or even the same species. For example, polyphenols related to catechol, which are produced by brown algae are known to inhibit the growth of many species of unicellular algae. Similarly toxins produced by the blooms of dinoflagellates (red tides) can lead to extensive mortality of fish. Typical average dissolved organic composition of sea water compiled from the existing data is given in Table 6.2.

Table 6.2: A Synopsis of the Average Dissolved Organic Composition of Seawater

Component	*Concentration in Seawater (as g.C.l^{-1})*
Vitamin B_{12}	0.0005
Thiamine	0.005
Biotin	0.001
Total fatty acids	5
Urea	5
Total free sugars	10
Total free amino acids	10
Total carbohydrates	200
Combined amino acids	50

Source: Williams, P.J. LeB, 1975.

6.1.3 Removal of DOM

DOM in sea water can be removed by three mechanisms (*a*) Physico-chemical removal arising from change of state, (*b*) chemical removal or modification, and (*c*) biological removal. Among them the biological mechanism predominates over others. The breakdown of DOM provides energy and cellular carbon for living organisms and brings about regeneration of carbon dioxide and nutrients.

Loss of DOM can result by transfer either to the solid or gaseous phase. Volatile compounds like methane, ethylene, propylene, carbon monoxide, dimethyl sulphide etc. may be produced during chemical and biological decomposition of organic material and may escape into the atmosphere. Losses may also occur as a result of conversion of DOM to POM by aggregation. Sorption on to the existing particle surfaces is often regarded as an important step in the removal of DOM. For example, proteins and polysaccharides are strongly adsorbed on natural detrital surfaces in sea water.

Substances like DDT, PCB' s, Vitamin B_{12} and biotin present in DOM undergo photochemical degradation. In sterile sea water, Vitamin B_{12} and biotin lose most of their activity while thiamine loses 50 per cent of its activity when exposed to sunlight by two weeks. On the other hand, the chemical decomposition (hydrolysis) of proteins is very slow and occurs over a period of hundred to thousands of years.

Degradation of DOM in the upper 100 m of oceans occurs predominantly by biological mechanism. Bacteria are the principal users and decomposers of detrital and dissolved organic matter in the sea. These micro organisms occur widely throughout the oceans where they are mainly associated with suspended particulate matter. In addition to heterotrophic bacteria, certain marine algae, small flagellates and soft bodied invertebrates also utilise DOM in sea water.

6.1.4 Distribution of DOM

We have seen earlier that most of DOM in the sea originates from the carbon dioxide fixed by phytoplankton. Therefore it would be expected that its concentration in the euphotic zone at a given location is approximately related to primary production which we will be discussing in Chapter 7. The actual concentration of DOM at any instant will depend on the balance between the rate at which DOM is formed and the rate at which it is removed by decomposition or utilization.

(a) Seasonal Variation

Seasonal variations of DOC are usually restricted to approximately upper 100 m and correlate roughly with productivity. The highest values are found in spring (March-June) and early summer (July-September) somewhat later than the period of maximum phytoplankton activity. The concentration then decreases slowly and remains constant, except some minor maximum in the autumn (September-December), till the onset of spring bloom. Seasonal variations of DOC in surface waters of open oceans in general, are much smaller than those in coastal waters.

(b) Vertical Variation

The depth distribution of DOC, DON and DOP in all the oceans follow the same pattern. Higher concentrations are generally found in surface layers. Below the euphotic zone, the concentrations begin to decrease with increasing depth depending upon the productivity, availability of heterotrophs and prevailing hydrographical conditions. At depths greater than few hundred metres, their concentrations remain practically constant since the DOM is resistant to decay. Typical profiles of dissolved and particulate organic nitrogen are shown in Figure 6.1. The distribution of both PON and DON show in general, higher concentrations at surface and lower values at depth. The near surface maximum in both these constituents are apparently related to the photosynthetic production. Distribution of organic phosphorus (POP and DOP) follows the same pattern of PON and DON. As expected high concentrations of POP occur in the productive upwelling zones. Dissolved organic phosphorus (DOP) accounts for 65-75 per cent of total phosphorus, with the rest occurring in particulate from (Naqvi, 2001).

The ratios of concentrations of DOC to DON in the nearshore surface waters lie in the range 100:15 to 100:25, approximately similar to that found in the phytoplankton. However, the values in deep waters vary considerably from one watermass to another. For example, the ratio of DOC to DON in deep waters off Southern California (100:8) differ widely from those of North Atlantic (100:50).

While considering the general distribution of DOC in the sea, anoxic environments merit special attention. We are aware that anoxic conditions prevail when oxygen consumption exceeds renewal in the water column, the best known examples are the Black Sea, Baltic Sea, Cariaco Trench and some Norwegian and Swedish fjords. In the absence of oxygen, the organic material undergoes decomposition by other oxygen donors such as nitrate, nitrite, sulphate and carbon dixodide. Eventhough the sea water contains comparatively small amounts of nitrate and nitrite, sulphate and carbon dixodide are present in much greater amounts. The vertical distribution of DOC in the Black Sea shows an increase with depth from about 2 mg C l^{-1} in aerated (oxic) zone to 6 mg C l^{-1} in the deep (anoxic) waters. This suggests that the organic matter in deep (anoxic) waters mostly comprises volatile products of bacterial fermentation

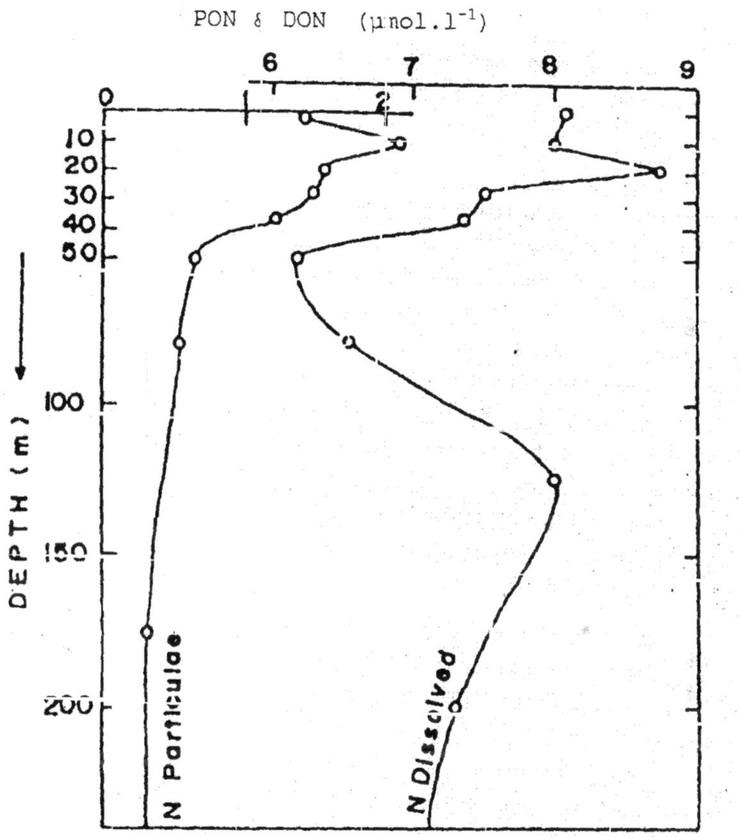

Figure 6.1: Vertical Profiles of Particulate Organic Nitrogen (PON) and Dissolved Organic Nitrogen (DON) in the Western Indian Ocean (*Source*: Naqvi, 2001)

such as methane which also shows pronounced increase with depth in the anoxic basins.

6.2.0 Particulate Organic Matter (POM)

Particulate organic matter (POM), sometimes called seston, is an important component of food chain in the sea. Information on its composition and distribution is essential as it provides food for organisms at several trophic levels. The POM includes both living and non-living suspended particles in the sea larger than 0.5 µm diameter. While phytoplankton constitutes a major fraction, bacteria, fungi, yeast, zooplankton constitute a minor portion of living matter. The non-living (detritus) fraction of POM essentially consists of the decay products of both plants and zooplankton. POM is commonly represented as particulate organic carbon (POC). The concentration of POC in open ocean waters vary in the range 20-200 µg C l^{-1}. Its concentration in coastal waters is generally one or two orders of magnitude higher than in oceanic waters which may be due to higher productivity in coastal waters and transport of allochthonous organic material from the land. In general, detrital fraction constitutes major portion of total POC, often exceeding the phytoplankton carbon by a factor of 10 or more except in large phytoplankton blooms in near surface waters.

In the determination of elemental composition of POM such as carbon (POC), nitrogen (PON) and phosphorus (POP), a known volume of sea water is filtered on a 0.5 µm glass fibre filter. The POC on the filter is digested with acid-dichromate solution and the amount of dichromate reduced is measured photometrically (Strickland and Parsons, 1968). The particulate nitrogen (PON) is determined by digesting the filter (Kjeldahl digestion method) with concentrated H_2SO_4 in presence of a catalyst (SeO_2). Ammonia so produced is determined photometrically by indophenol blue method as described earlier (4.1.0). A C-H-N analyzer can be employed for determination of carbon and nitrogen. Particulate phosphorus (POP) is determined by digesting the filter containing POM with perchloric acid. The orthophosphate so formed is then determined photometrically by molybdenum blue method (4.2.0). In addition to the above elements, photometric methods are available for the determination of certain classes of organic compounds such as plant pigments, carbohydrates, lipids and amino acids.

6.2.1 Sources of POM

Main sources of POM in the sea are (a) materials of terrestrial origin, transported to the sea by rivers or through atmosphere (allochthonous material) and (b) substances present (autochthonous) in the marine environment. These include (i) fragments of plants and animals derived from marine food chain, and (ii) POC formed *in situ* through a complex equilibrium between DOC and POC.

As we have discussed earlier, certain substances like DDT and its derivatives are transported into the sea through atmosphere by winds. Particles carried by rivers influence most nearshore environments. In contrast, POC in the oceanic waters is produced *in situ* by phytoplankton. This is evident from the fact that high concentrations of POC in the surface layers of oceans are generally associated with areas of high primary production. As stated earlier, POC can be formed from DOC, either through bacterial activity or through its adsorption on organic particulate surfaces. Similarly DOC can be produced through decomposition of POC and solution processes. Thus the inter conversion of DOC and POC in the sea can be regarded as a reversible process.

POM in sea water can be approximately subdivided according to size. The smallest particles (greater than 0.5 μm upto few tens of μm) comprises of bacterial and algal cells, fine organic detritus (coccoliths and diatom skeletons). The medium size (tens to few hundreds of μm) particles are represented by large detritus and faecal pellets which are produced by biological aggregation or packaging. The larger size aggregates (several millimeter to centimeters) commonly known as marine snow (or fluff) consist of detritus, living organisms (including bacteria) and some inorganic matter (clay minerals).

Marine snow aggregates provide a mini ecosystem within which the processes of photosynthesis, decomposition, and nutrient regeneration may occur at rates greater than in surrounding waters. The marine snow is most abundant in surface waters where production is high and the component particles are available in plenty. The chemical and biological properties of these particles change on time scales of hours to days as the microbial communities undergoes complex successional changes. Since the biological ' glues' that bind the aggregates are not very strong, marine snow undergo continuous disintegration and aggregation. The breakdown of marine snow is accomplished by decomposition (dissolution of components and/or the biological glues); disaggregating as a result of turbulence; and consumption by animals. The aggregates are lost from their place of formation by vertical settling and/or by horizontal advection by currents. Most of the marine snow is disaggregated and/or eaten and repacked into faecal pellets in the upper 500-1000 m of the water column. However, faecal pellets (100-300 μm size) provide the main component of the sinking particle flux below the depth of 1000 m.

POM also occurs as surface-active films on the sea surface over extensive areas or as sea-surface microlayers. The films which are monolayer of thickness are composed of fatty acids (C_8–C_{18}), methyl esters of fatty acids (C_{11} – C_{22}), higher aliphatic alcohols (C_{12}, C_{16} and C_{18}) and hydrocarbons. These film forming materials originate from the metabolites of phytoplankton and zooplankton gathered on the bubbles which carry them to the surface and disperse them when they burst. The films often modify the physical properties of sea surface including surface tension and viscosity, and lead to the production of slicks in the areas of wave damping.

6.2.2 Nature and composition of POM

Live phytoplankton which comprises bulk of the biomass in the sea is confined to the euphotic zone and the water lying immediately beneath it. The chemical composition of phytoplankton (consisting of several classes such as Chlorophyceae, Chrysophyceae, Bacillariophyceae, Dinophyceae, Myxophyceae etc) varies from species to species and according to environmental conditions. One important chemical property of POM is carbon to nitrogen ratio. In deep water samples the ratio lies in the range 10-20 while its value is about 6 in the healthy phytoplankton of euphotic zone. It is apparent that there is a loss of nitrogen from below the euphotic zone.

The principal amino acids of phytoplankton (proteins) are glutamic acid, aspartic acid, alanine and leucine together with lesser amounts of about 20 others. Below 200 m depth, the proportion of serine, lycine and arginine increase while that of alanine decrease. The usual plant pigments of phytoplankton consist of carotene, chlorophylls and xanthophylls. Concentration of chlorophyll- decreases with depth because of its decomposition to phaeo-pigments. However, carotenoids are less readily decomposed than chlorophylls. The carbohydrates of phytoplankton occur in three principal forms (*a*) as polysaccharide components of the cell wall which are often resistant to bacterial decay, (*b*) as mono and oligosaccharides in the cell sap, and (*c*) as polysaccharides which serve as food reserves. It has been observed that the water soluble fraction of phytoplankton carbohydrates decreases with depth and they disappear in the depth range 300-1000 m, leaving only water insoluble fraction. In addition to these major metabolites, phytoplankton contain small amounts of other biologically important compounds, such as vitamins, nucleic acids etc. These substances may play an important part in the food chain as many of them can be utilised directly by zooplankton to satisfy their essential requirements.

A major portion of nonliving POM (detritus) comprises of the decay products of organisms such as cell walls of the phytoplankton and the chitinous exoskeletons of zooplankton. The composition of marine detritus varies considerably with depth. In the euphotic zone it consists to a large extent, major biochemical metabolites (such as carbohydrates, proteins, plant pigments) and their degradation products. Many of them are readily oxidised by bacteria and they are subjected to predation by filter feeders. Probably only a small fraction of resistant parts of organism, such as the cell walls, the exoskeletons of zooplankton and the precipitated humic substances will survive at higher depths (100-200 m). While phytoplankton (living) are abundant, detritus probably acts only as a supplement to the diet of zooplankton. However, when the crop is sparse, it may form a major part of the food of these animals. Bacteria use the particulate matter to supply both their energy and the materials for their protoplasm. During their respiration and metabolic processes, CO_2, ammonium and phosphate ions are regenerated.

Particulate organic matter in the oceans is generally associated with large (40-60 per cent) inorganic material, which consists of silica, ferric oxide, alumina, calcium and carbonates. In coastal environments where there is heavy runoff, the inorganic fraction largely consists of local terrestrial (clay and carbonate) minerals.

6.2.3 Distribution of POM

POC in a column of sea water is always higher (10 times) than the (yearly) primary productivity which will be discussed in Chapter-7. The POC concentration in the euphotic zone is relatively high and variable which can be roughly correlated with phytoplankton activity. For this reason POC exhibits spatial and temporal variations similar to primary productivity. A typical vertical profile of POC in the Bay of Bengal is shown in Figure 6.2a. Higher POC concentrations occur between 40-60 m due to the combined

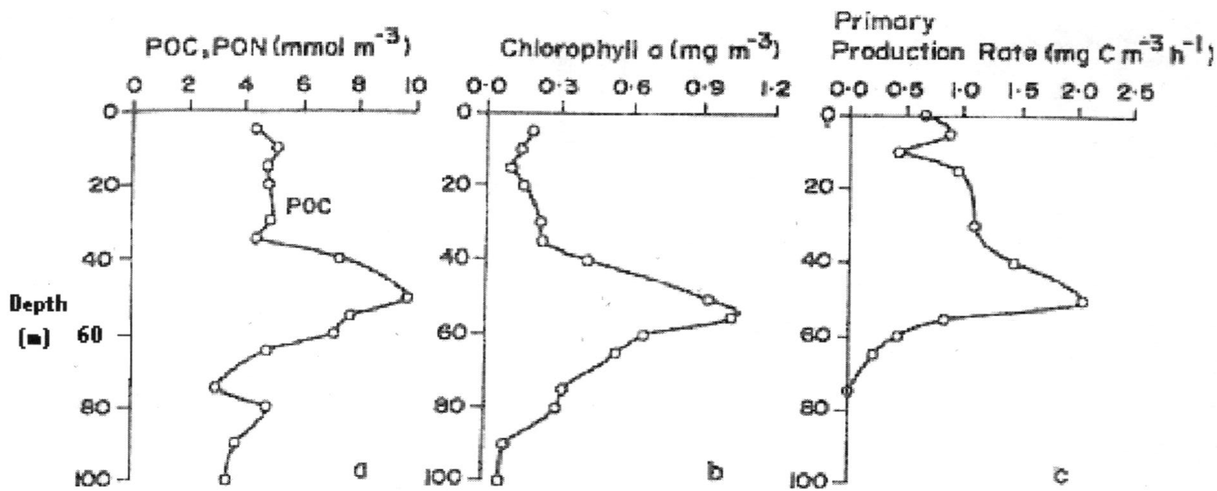

Figure 6.2: Vertical Profiles (a) Particulate Organic Carbon (POC); (b) Chlorophyll-a and b and (c) Primary Production in the Bay of Bengal
(*Source*: Naqvi, 2001)

effect of biological activity and huge inputs of allochthonous organic matter in the Bay (Naqvi, 2001). The decrease of POC with depth is attributed to zooplankton grazing (ca 30 per cent) and to degradation processes including bacterial decomposition. Vertical distribution of POC closely resembles that of the chlorophyll-**a** and primary production (Figure 6.2 b and c) in the region.

Beneath the euphotic zone the concentration of POC (both living phytoplankton and detritus) decreases rapidly below depth of 200 m and then remains more or less constant upto the bottom. The decrease of POC in the depth interval 0-100 m can be given by an equation:

$$C = C_o e^{-k\,(Z-Zo)} \quad 6.1$$

where,

C is the concentration of carbon at depth Z and C_o is the concentration at depth Z_o. In the depth range 0-30 m, the k value vary from 0.019-0.039, the highest value of k being associated with the highest value of C_o.

Multiple Choice Questions

1. The average concentration of dissolved organic matter (DOC) in the ocean is

   (a) 0.5–2.0 µg C 1^{-1} (b) 20–200 µg C 1^{-1}

   (c) 0.5–2.0 mg C 1^{-1} (d) 10–20 mg C 1^{-1} ( )

2. The C/N ratio of POM in deep sea water

   (a) Is greater than in phytoplankton

   (b) Is less than in phytoplankton

   (c) Is less than in sediments

   (d) Is greater than in surface water ( )

3. Concentration of DOM and POM in the euphotic zone in general, are

   (a) Higher than the deeper waters

   (b) Lower than the deeper waters

   (c) Nearly equal to the deeper waters

   (d) Not comparable ( )

4. The vertical profile of POC indicates
   (a) An increase with depth
   (b) Decrease with depth
   (c) No change with depth
   (d) Increase followed by decrease with depth ( )

5. The concentration of chlorophyll- in sea water
   (a) Increases with depth
   (b) Remains same in the euphotic zone
   (c) Decreases with depth
   (d) Increases upto euphotic zone and then decreases with depth ( )

6. The non-living fraction of POM is called as
   (a) Seston (b) Neuston
   (c) Detritus (d) None of the above ( )

Short Answer Questions

1. How do you distinguish dissolved and particulate matter in the sea?
2. Broadly indicate the chemical composition of DOM.
3. What is Gelbstoff? Indicate its composition.
4. What are the important processes of removal of DOM?
5. Indicate the sources of POM in sea water.
6. Explain the removal mechanism of POM from sea water.
7. What is the mechanism of conversion of DOM to POM and viceversa.
8. How do you quantitatively explain the decrease of POC within 0-100m depth range?
9. What is marine snow? Indicate its significance.
10. What do you understand by the term detritus?

Review Questions

1. Discuss in detail the sources, nature and chemical composition of dissolved organic matter (DOM) in the sea.
2. Explain the distribution of DOC and POC in the sea.
3. Give an account of sources, nature and chemical composition of particulate organic matter (POM) in the sea.

Answers to

Multiple Choice Questions

1. c 2. a 3. a 4. b 5. c 6. c.

Chapter 7

Photosynthetic Productivity

Photosynthesis is regarded as the most important stage in the marine food chain. During this process, plants (particularly phytoplankton) remove dissolved carbondioxide and nutrients from sea water, and using solar energy, convert them to complex organic molecules. Thus the light energy is converted into chemical energy and stored in plankton. Much of the organic carbon fixed by plants in this way is consumed by organisms at higher trophic levels (zooplankton, fish etc.), by breaking down the complex organic molecules and thus releasing the stored chemical energy. This process is known as respiration. Thus, photosynthesis and respiration are chemically opposite and complementary processes, which maintain the delicate balance of the life cycle. Chemical reactions involved in these two processes can be simplified as follows:

$$\text{Photosynthesis: } H_2O + CO_2 \xrightarrow[\text{Chlorophyll}]{\text{Sunlight (h\ \ )}} CH_2O + O_2 \qquad 7.1$$

$$\text{Respiration: } CH_2O + O_2 \xrightarrow{\text{Enzymes}} CO_2 + H_2O \qquad 7.2$$

However, these processes are highly complex in their nature and only a brief account of them are given here.

7.1.0 Mechanism of Photosynthesis

The basic mechanism of photosynthesis can be envisaged as a three stage process namely:

(i) Generation of Reducing Power Sunlight (h)

$$H_2O + H_2O \xrightarrow[\text{Chlorophyll-}]{\text{Sunlight (h)}} O_2 + 4H^+ + 4e^- \qquad 7.3$$

Thus the H^+ ions are produced from water in photosynthesis. Chlorophyll-**a** not only absorbs photons from sun light but also transfers this trapped energy to the water molecules.

(ii) (a) Storage of Energy

$$4H^+ + 4e^- + ADP + Pi + (O_2) \longrightarrow 2H_2O + ATP \qquad 7.4$$

where,

ADP and ATP are adenosine di and triphosphates, respectively, Pi is inorganic phosphate and O_2 is oxygen bound to NO_3 or NO_2. Combining eqs. 7.3 and 7.4, we get:

$$ADP + Pi \longrightarrow ATP \qquad 7.5$$

(ii)(b) Storage of Reducing Power

$$2H^+ + 2e^- + NAD \longrightarrow NADH_2 \qquad 7.6$$

where,

NAD is nicotinamide-adenosine dinucleotide, and $NADH_2$ is its reduced form. Combination of eqs. 7.5 and 7.6 after balancing the terms and rearranging, gives:

$$3\ ATP + 2\ NADH_2 \longrightarrow 3\ ADP + 2\ NAD + 3\ Pi + 4\ H^+ + 4\ e^- \qquad 7.7$$

(iii) Reduction of CO_2 to Carbohydrates

$$CO_2 + 4\ H^+ + 4\ e^- \longrightarrow CH_2O + H_2O \qquad 7.8$$

Combining eqs. 7.7 and 7.8 results in the overall reaction:

$$CO_2 + 2NADH_2 + 3ATP \longrightarrow CH_2O + H_2O + 3ADP + 3Pi + 2NAD \quad 7.9$$

It is important to note that light energy is involved only in the first reaction (eq.7.3), namely, the generation of reducing power. For convenience, we can subdivide photosynthesis by green plants into two basic reactions, namely:

1. Light reactions concerned with conversion of light energy to chemical energy in the form of ATP and $NADH_2$ (eqs.7.3-7.6), and
2. Dark reactions concerned with utilisation of trapped energy for the conversion of CO_2 to complex organic molecules (eqs.7.7–7.9).

Both light and dark reactions take place within the chloroplasts of eukaryotic algae, where chlorophyll-**a** and accessory pigments are usually found. The light required by plankton for photosynthesis lies in the visible region of the electromagnetic spectrum (400-700 nm). This is also absorbed by chlorophyll-**a** which has maximum absorption in 650-700 nm range.

7.1.1 Chemosynthesis

Some bacteria are capable of generating reducing power. The hydrogen donor in this case is never water, but usually H_2S or CH_4. These bacteria are anaerobic and can live only in anoxic environments and never liberate oxygen. The process of organic production by anaerobic bacteria is called chemosynthesis. In chemosynthesis the first stage is represented as:

$$XH_2 + H_2O \xrightarrow[\text{(enzyme)}]{\text{dehydrogenase}} XO + 4H^+ + 4e^- \quad 7.10$$

The rest of the stages (*ii*) and (*iii*) are similar to that of photosynthesis (eq.7.4–7.9) except that the oxygen (O_2) is made available from SO_4 rather than NO_2 or NO_3 which cannot exist under anoxic conditions. In anoxic environments such as the Black Sea, the Baltic Sea, the Cariaco Trench and in some Norwegian and

Swedish Fjords, chemosynthetic production in the water column is quite significant. However, in most pelagic regions, the amount of CO_2 fixed by chemosynthesis rarely exceeds 5 per cent of that fixed by photosynthesis.

7.1.2 Respiration

Energy required by the algae for their metabolic processes is obtained by the oxidation of photosynthetically produced complex molecules to compounds of lower (potential) energy, and ultimately to CO_2. This process known as respiration, is similar to the simplified reaction pathway represented by eq.7.2. Respiration takes place both in the light and in the dark with almost at the same rate.

7.2.0 Phytoplankton Production and its rate Measurement

In order to quantify phytoplankton potentially available to food chain in a particular location, we must take into consideration not only the amount of phytoplankton available at any instant, but also the rate at which it is produced. The former is termed as the " Standing Crop" and is usually expressed as the amount of living phytoplankton (as mg of carbon) present in 1 m^3 of sea water or beneath 1 m^2 of sea water. The rate of photosynthetic production is usually defined as the weight of inorganic carbon fixed photosynthetically in unit time per unit volume or unit area of surface. It is generally represented as mg C m^{-3} h^{-1} or mg C m^{-2} d^{-1}. It is necessary to distinguish between the gross rate of production and the net rate of production. The former is the rate at which the plant carbon is produced photosynthetically while the latter is the gross rate minus the rate of loss of carbon through respiration by plants.

A consideration of the general equation of photosynthesis (eq.7.2) suggests that the gross rate of photosynthesis can be determined by measuring (*i*) the rate of liberation of oxygen, (*ii*) the rate of removal of CO_2, and (*iii*) the rate at which the inorganic carbon is converted to organic matter (plant tissue). Among them (*i*) and (*ii*) are more commonly used since there are no satisfactory methods for the direct measurement of low concentrations of plant material or organic matter produced during photosynthesis in the sea. Let us consider the principles of these two methods briefly.

7.2.1 Measurement Based on Oxygen Liberation

The light and the dark bottle method measures organic production in the sea in terms of changes in oxygen concentration (details are available in Strickland and Parsons, 1968). In this method, sea water samples are collected into 300 ml BOD glass stoppered bottles (in triplicate) at each depth. The dissolved oxygen content of sea water is then determined by Winkler' s method (3.2.0) in one bottle immediately after collection. The other two bottles, one light (ordinary) and the second black (wrapped with a black paper or painted) are suspended in the water column at different depths throughout the euphotic zone by affixing them onto a winch wire. After a suitable period of incubation (3-8 hrs), the dissolved oxygen in the dark and light bottles is again determined. The increase in the oxygen in light bottle is a measure of the mean gross photosynthesis which has taken place. The loss of oxygen which occurs simultaneously due to respiration of plant cells is assessed from the decrease in its content in the dark bottle. The method is, however, unsuitable for waters of low productivity (oligotrophic) and polluted eutrophic waters (particularly if bacterial populations are high).

7.2.2 Measurement Based on Uptake of CO_2

This method based on the use of radioactive carbon-14 isotope is suitable for measuring photosynthesis rate both in oligotrophic and eutrophic waters. In this method, a series of glass stoppered 300 ml BOD bottles filled with sea water are treated with 2 ml of sodium bicarbonate solution in which the carbon is labelled with C-14 (1-25 μCi). The bottles are then suspended in the water column at different depths and incubated for a period of 3-8 hours. The contents are then filtered through a 0.45 μm membrane filter paper, the filter is washed with sea water and dried at room temperature. The amount of carbon-14 fixed is then measured on a Scintillation or Geiger-Muller Counter. In order to compensate for CO_2 uptake by non-photosynthetic organism, a blank is carried simultaneously using a dark bottle as described earlier in oxygen method. The photosynthetic productivity is then calculated using the equation:

$$\text{Productivity (mg C m}^{-3}\text{ h}^{-1}) = \frac{1.05\,(C_S - C_D)\,W}{C.\,N} \qquad 7.11$$

where,

W: Weight of total CO_2 content in sea water ($mgCm^{-3}$) which can be obtained from alkalinity and pH of sea water.

N: Number of hours of incubation of sea water

C_S and C_D: Normalised counting rates of light and dark bottles respectively,

C: Normalised counting rate of C-14 which was originally added, and 1.05 is a factor.

Although this method is most satisfactory, it also suffers from certain limitations. They include (*a*) uncertainity as to whether it measures gross or net production, (*b*) it measures only fixation of inorganic carbon and does not take into account any photo assimilation of inorganic compounds, and (*c*) photosynthesis may give rise to soluble extra cellular metabolites (*e.g.* glycollic acid) which will not be retained on the filter paper. Thus the productivity will tend to be underestimated.

7.3.0 Factors Controlling the Primary Production

Important factors controlling primary production in the oceans are (*i*) light, (*ii*) temperature, (*iii*) salinity, (*iv*) nutrients and trace elements, and (*iv*) organic compounds

7.3.1 Light

As the energy source for photosynthesis is light, it is obvious that the quality and quantity of ambient light will influence photosynthetic production in the sea. The influence of light intensity on photosynthesis (shown in Figure 7.1) indicates three general features: (*a*) The linear phase where the photosynthetic production increases with light intensity, (*b*) The saturation plateau where the photosynthetic rate reaches its maximum, and (*c*) photo inhibition phase where photosynthetic production decreases with increase of intensity of light.

In temperate regions, the light energy (average over day and night) expressed as power(watts) that required to maintain the balance between production and respiration is in the order of 1.5-6.3 W m^{-2}. The photo inhibition which varies from one algal species to another in general, occurs at values ranging from 25-150 W m^{-2}. On the other hand, the intensity of sun light at the surface of sea

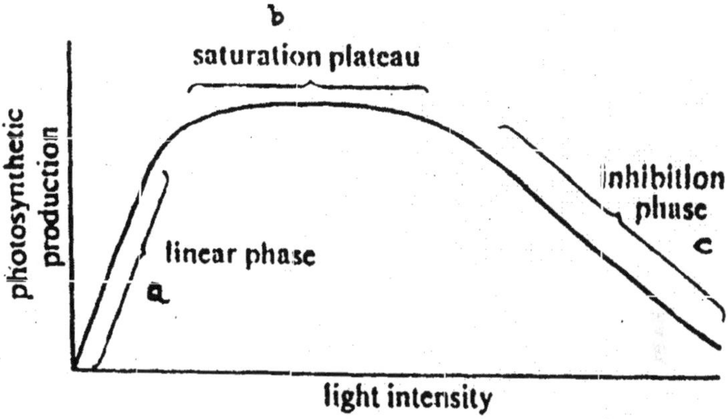

Figure 7.1: The Relationship Between Light Intensity and Photosynthetic Production (*Source*: Open University, 1978a)

water is much higher, of the order of 1000 W m^{-2}. In fact, it appears that about one quarter of the incident solar radiation is absorbed within 3 mm of the surface. Hence only specialised neuston can survive in these topmost few millimeters since they adopt themselves to the high intensity of radiation.

The intensity of radiation at surface as well as in different depths of the photic zone depends on several factors such as latitude, time of the day, weather conditions and the clarity of the water which determines the degree of scattering and selective absorption of different wave lengths by the sesten. In general, the effect of photo inhibition near the surface and photo limitation at depth combine to produce a distribution of primary production in the oceans as depicted in Figure 7.2. The photosynthetic production gradually increases from surface and reaches maximum at about 20-25 m and then decreases upto 100 m depth. Even in clean tropical oceans, 1 per cent of the surface visible light energy penetrate only to a depth of around 100 m, and thus limiting the production to a minimum.

The upper layer of the sea in which the gross primary production exceeds respiration is termed as the euphotic zone. The depth at which the net production is zero is termed as the compensation depth. It is defined as the depth at which the oxygen consumed by plants in respiration, over a period of 24 hrs, equals the amount produced by photosynthesis.

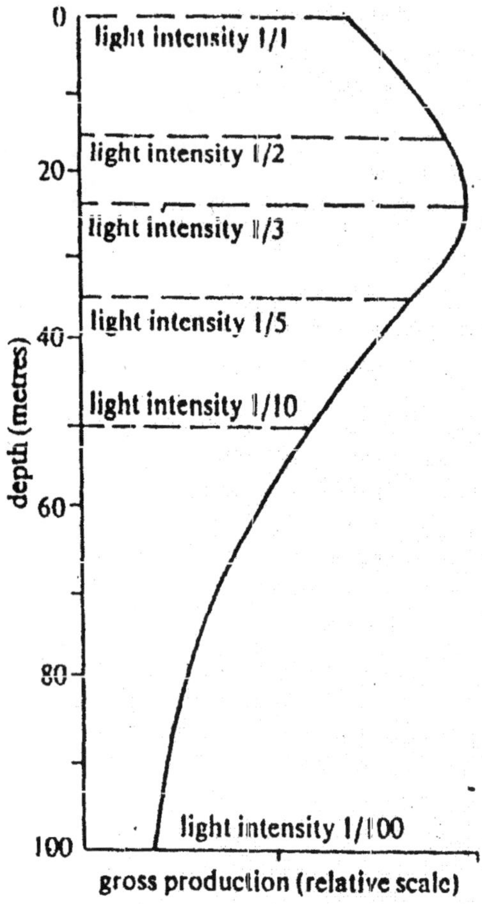

Figure 7.2: Gross Primary Production at Different Depths, with Light Intensities Expressed as Fraction of the Intensity at the Sea Surface (*Source*: Friedrich, 1969)

7.3.2 Temperature

Light reactions (eqs.7.3-7.6) of photosynthesis are comparatively independent of temperature, but the dark reaction (eqs.7.7 – 7.9) are dependent. Very high temperatures inhibit photosynthesis, presumably by damaging the enzymes and the cell structure as in photo-inhibition. In general, photosynthetic production occurs efficiently over the whole range of temperatures encountered in the marine environment (0– ~25^0 C). However, phytoplankton will tolerate only limited changes of temperature and are rapidly killed at temperatures of 10-15^0 C above that at which they are adopted to live.

7.3.3 Salinity

Although variations in salinity have some effect on the rate of photosynthesis, the optimum lies in the range 15-20 psu. Many of the phytoplankton species grow best at low salinities, thus the optimum salinity for the growth of *skeletonema* is 15-20 psu. Some species, particularly diatoms, grow poorly at salinities greater than 35 psu, and this perhaps explain their preference for coastal waters. The influence of small variations in salinities in oceans on productivity is quite insignificant. However, in estuaries, salinity rather than temperature appears to control productivity.

7.3.4 Nutrients and Trace Metals

For the healthy growth, phytoplankton require supply of nitrogen and phosphorus. Defficiency of nutrients particularly nitrate probably is the main factor limiting marine primary production in the oceans. However, the pollution in restricted water bodies such as estuaries, lagoons etc. with large amounts of nitrogen and phosphorus can lead to heavy blooms of phytoplankton. This process is known as eutrophication. Silicon is essential for the growth of those organisms such as diatoms and silicoflagellates, which possess siliceous frustules or skeletons. However, silicon is not a limiting element for the production of these organisms.

It is well known that plankton require for their healthy growth, a number of trace metals such as Fe, Mn, Cu, Co, Zn, Mo and V in addition to nutrients. The role of many of these elements in algal metabolisms have now been established. Thus iron as non-haem protein complex, called ferredoxin takes an essential part in the light reaction of photosynthesis. Manganese is known to be present in the enzyme cofactors involved in photosynthesis, and also in nitrate reduction, in which cofactors containing molybdenum are important. Copper-complexes serve as cofactors in oxidation–reduction cycles.

7.3.5 Organic Compounds

In addition to nutrients and trace metals, certain organic compounds such as vitamin B_{12}, thiamine (B_1), biotin, ascorbic acid and cystine act as growth promoters for phytoplankton and algae. There is considerable evidence for the existence of large healthy populations of photosynthetic organisms at depths far below the euphotic zone in some parts of the oceans. It seems likely that these algae live heterotrophically on the dissolved organic compounds present in sea water.

7.4.0 Seasonal Variation of Primary Production

Away from the immediate vicinity of coasts (where factors such as nutrients derived from land drainage, tidal and current mixing, upwelling, low salinity and high turbidity complicate the situation), seasonal variation in algal biomass and primary production generally correlate with latitudes. Variation in primary production is most marked at high latitudes for example, in the Arctic or Antarctic, where sufficient light for photosynthesis is only available during five or six months in a year. However, because of absence of thermocline, effective mixing of water column brings abundant nutrients into the photic zone, so that during the 3 or 4 months when net photosynthesis is possible, very large algal bloom is quickly produced (Figure 7.3a). The response of herbivores to the bloom has a time lag of about 6 weeks, so the algal biomass continues to increase until reduction in light intensities and the onset of peak grazing pressure combine to end the bloom.

On the other hand, moving towards low latitudes (equator), the amplitude of these seasonal fluctuations decreases, resulting in only small variations in algal production (Figure 7.3b). At these low latitudes, light intensity is very high throughout the year, but the steep permanent thermocline ensures vertical stability and prevents mixing between surface and deep layers resulting in permanent shortage of nutrients over wide areas. Hence algal production and herbivore biomass are comparatively constant, and grazing pressure is intense throughout the year. Primary production is thus limited by grazing or lack of nutrients or both.

7.5.0 Variation of Primary Productivity in the Indian Ocean

The productivity in the Indian Ocean varies temporarily and spatially. This is evident from Figure 7.4A and B where a number of regions of high and low productivity can be recognised. A substantial

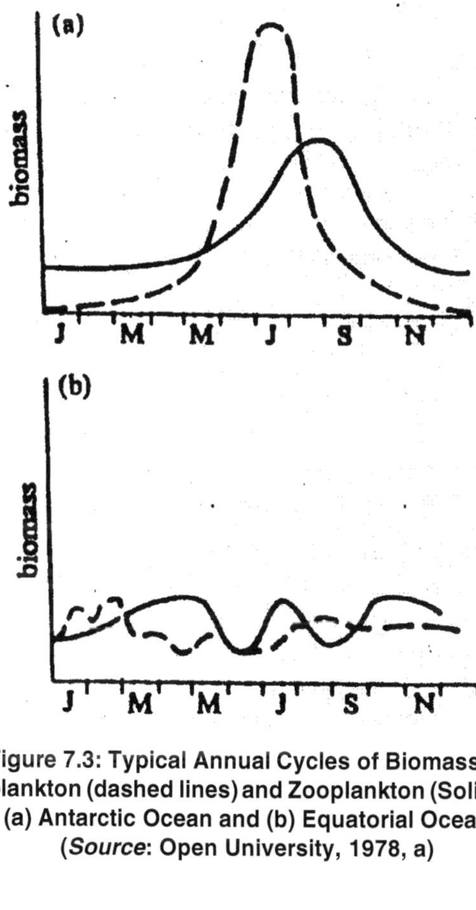

Figure 7.3: Typical Annual Cycles of Biomass of Phytoplankton (dashed lines) and Zooplankton (Solid Lines). (a) Antarctic Ocean and (b) Equatorial Ocean (*Source*: Open University, 1978, a)

portion of northern Indian Ocean (Arabian Sea) is highly productive during southwest monsoon (Figure 7.4A). Upwelling in the Somali region lying 5°–10° N latitudes starts in May and peaks up by July-August resulting in high primary productivity (1.3–1.8 g. C. m^{-2} day^{-1}). Between 55°–65° longitude the rich areas extend upto south of 15° S along the coast from east Africa to south Africa. Extensive areas of high productivity (0.75 g.C. m^{-2} day^{-1}) are located along the east coast of Arabia and northwest India. In the eastern Arabian Sea, surface production of 0.23 g.C.m^{-2} day^{-1} occurs along the Southwest of India during upwelling period (S.W. monsoon). In general, areas of low productivity were more extensive during the

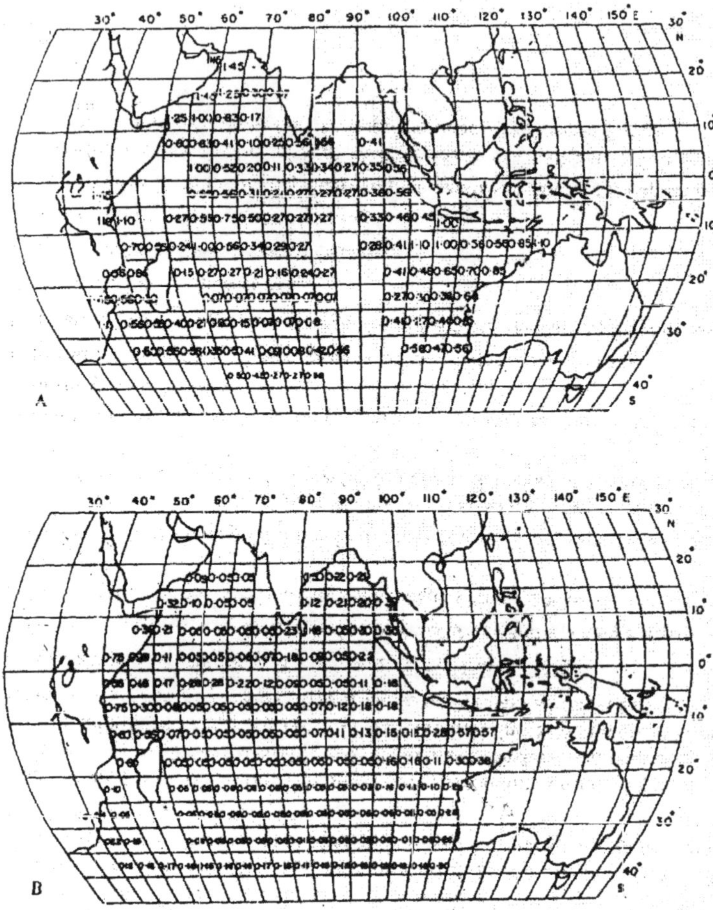

Figure 7.4: Primary Productivity of the Indian Ocean in gC m^{-2} day^{1}. (A) In the southwest monsoon; (B) in the northeast monsoon (*Source*: Verlecar and Parulekar, 2001)

N.E. rather than S.W monsoon seasons (Verlecar and Parulekar, 2001).

Compared to the Arabian Sea, the Bay of Bengal waters are less productive. While nutrient inputs from many major rivers are large, the narrow shelf, heavy cloud cover and reduced light penetration limit the production rate. The Bay of Bengal is also influenced by localized upwelling along the east coast of India during S.W monsoon. Production is high off Sri Lanka, off Java, west and northwest coasts of Australia. Unlike Arabian Sea, the surface production along the east coast of India and eastern Indian ocean are high during N.E. monsoon (0.1–0.2 mg.C.m^{-2} day^{-1}).

Open waters of central Indian Ocean on the north and south of equator have the lowest primary productivity (0.05–0.06 mg.C.m^{-2} day^{-1}). Similar areas called " desert regions" are known to exist in the Atlantic and the Pacific Oceans.

Multiple Choice Questions

1. Standing crop is expressed by the units
   (a) mg C m^{-3} (b) mg C m^{-3} d^{-1}
   (c) mg C m^{-2} d^{-1} (d) mg C m^{-3} d^{-1} ( )

2. Primary productivity, in general, is expressed as
   (a) mg C m^{-3} (b) mg C m^{-2}
   (c) mg C m^{-3} d^{-1} (d) None of them ( )

3. Compensation zone is that region in the sea where
   (a) Primary productivity exceeds respiration
   (b) Respiration exceeds primary productivity
   (c) There is no primary productivity
   (d) There is no net production ( )

4. Photo inhibition of phytoplankton occurs at
   (a) High salinities
   (b) High nutrient concentrations
   (c) High light intensities
   (d) None of the above ( )

5. Primary production, in general, is most marked at
   (a) Equatorial region
   (b) High latitudes
   (c) Low latitudes
   (d) Regions of high vertical stability ( )

Short Answer Questions

1. Explain the basic mechanism of photosynthesis.
2. Explain the principle and method of estimation of primary productivity using Carbon-14.
3. What are the factors affecting primary production in the sea?
4. Explain the geographical distribution of primary production.
5. How the spatial and temporal variation of phytoplankton correlate with POC?

Review Questions

1. Define the terms standing crop and primary productivity in the sea. Explain the methods for the determination of primary productivity and their limitations.
2. What do you understand by the terms photosynthesis and chemosynthesis? Explain their mechanism and significance in the sea.
3. Describe the seasonal and geographical variation of primary production in the oceans.
4. Distinguish the terms Photosynthesis and Respiration. Describe the factors which influence primary production in the sea.

Answers to

Multiple Choice Questions

1. a 2. c 3. d 4. c 5. b.

Bibliography

Andersson, L. and D. Dyrssen, 1994. Alkalinity and total carbonate in the Arabian Sea. Carbonate depletion in the Red Sea and Persian Gulf. Marine Chemistry, 47, 195-202.

Anonymous 1954. Andhra University Memoirs in Oceanography, Vol.I, Andhra University Series No. 49, 162pp.

Anonymous 1958. Andhra University Memoirs in Oceanography, Vol.II, Andhra University Series No.62, 237pp.

Aston, S.R., 1980. Nutrients, dissolved gases and general biogeochemistry in estuaries, In: Chemistry and biogeochemistry of estuaries, Olausson, E. and Cato, I.(Eds.) John Wiley. New York, 233-262.

Boyle, E. and Edmond, J.M. 1975. Copper in surface waters South of New Zealand, Nature; 253: 107-9.

Brewer, P.G. 1975. Minor elements in Sea water, In: Chemical Oceanography, Vol.I, (2nd Edn.) Riley, J.P. and Skirrow G.(Eds), Academic Press, London, 415-496.

Broecker. W.S. 1974. Chemical Oceanography, Harcourt Brace Joranovich, New York, 214pp.

Burton, J.D. 1975. Radioactive nuclides in the marine environment, In: Chemical Oceanography, Volume 3, (2nd Edn.). Riley, J.P. and Skirrow, G.(Eds.), Academic Press, London, 91-191.

Dileepkumar, M., Rajendran, A., Somasundar, K., Ittekot, Y and Desai, B.N., 1992. Processes controlling carbon components in the Arabian Sea, In: Oceanography of the Indian Ocean. Desai, B.N., (Ed.), Oxford and IBH Pub., New Delhi, 313-325.

Dileepkumar, M., Naqvi, S.W.A., Jayakumar, D.A., George, M.D., Narvekar, P.V. and Desousa, S.N., 1995. Carbondioxide and nitrous oxide in the north Indian Ocean, Curr. Sci., 69, 672-678.

Dittmar, W., 1884. Physics and Chemistry. Rept. Sci. Res. H.M.S. Challenger, 1873-76, 1,251 pp.

FAO, 1971. Report on the Seminar on methods of detection, monitoring and measurement of Pollutants in the Marine Environment, FAO Fisheris Rep. No.99, Spl.1 Food and Agricultural Organisation of United Nations, Rome.

Friedrich, H. 1969. Marine Biology, Sidgwick and Jackson (Pub.), London, 474 pp.

Horne, R.A. 1969. Marine Chemistry, Wiley Interscience, New York, 568pp.

IIOE, 1965-72. International Indian Ocean Expedition Collected Reprints, Vols. I – III, UNESCO, Paris.

IIOE, 1969-73. IIOE Zooplankton Atlases, 1-5 and hand books 1-5. Indian Ocean Biological Centre, National Institute of Oceanography, Cochin.

Krey, J and Babenerd, B., 1976. Phytoplantkon production Atlas of the International Indian Ocean Expedition, UNESCO, Paris, 70 pp.

Liss, P.S. and Slater, P.G. 1974. Flow of gases across the air-sea interface, Nature, 247: 181-184.

Martin, Dean, F., 1970. Marine Chemistry, Vol.2: Theory and Applications, Marcel Dekker Inc., New York, 451 pp.

Murty, C.S. and Murthy, V.S.N., 2001. Physical Oceanography, In: The Indian Ocean – a perspective. Sen Gupta, R and Desa, E. (Eds.), Vol.I, Oxford and IBH, New Delhi, 1-59.

Nandakumar, K., Bhosle, N.B. and Wagh, A.B. 1987. Dissolved and particulate lipids in the Arabian Sea off Bombay Coast, Indian J. Mar. Sci., 16: 240-242.

Naqvi, S.W.A., Desousa, S.N., Fondekar, S.P. and Reddy, C.V.G., 1979. Distribution of dissolved oxygen in the western Bay of Bengal. Mahasagar, Bull. Natn. Inst. Oceanography. 12; 25-34.

Naqvi, S.W.A. and Qasim, S.Z. 1983. Inorganic nitrogen and nitrate reduction in the Arabian sea, Indian J. Mar. Sci., 12: 21-26.

Naqvi, S.W.A., 2001. Chemical Oceanography. In: The Indian Ocean – a perspective. Sen Gupta, R. and Desa, E. (Eds.). Vol.I, Oxford and IBH, New Delhi, 159-236.

Open University, 1978. Chemical Processes, Open University course material on Oceanography, Units 7 and 8. Wright, J.B.(Ed.) The Open University Press, Milton, Keynes, (U.K)., 37 and 54 pp.

Open University, 1978a. Biological Environments; Open University course material on Oceanography, Unit 9, Wright, J.B.(Ed.), The Open University Press, Milton, Keynes, U.K., 60 pp.

Open University, 1989. Sea water: its composition, properties and behaviour, Vol.2, Open University course material on Oceanography, Gerry Bearman (Ed.). Pergamon/Open University Press, U.K., 165 pp.

Open University, 1995. Sea water: its composition properties and behaviour, Vol.2, Open University course material on Oceanography, Garry Bearman (Ed.). Pergamon/Open University Press, 2nd Edition. U.K. 168 pp.

Open University, 1995a. Ocean Chemistry and deep sea sediments, Vol.5, Open University course material on Oceanography, Gerry Bearman (Ed.). Pergamon/Open University Press, 134 pp.

Picciotto, E.E. 1961. Geochemistry of radioactive elements in the ocean and the chronology of deep sea sediments, In: Oceanography, Sears, M(Ed.), Am.Assoc. Advan. Sci. Pub.No. 67. Washington D.C, 367-390.

Qasim, S.Z., 1977. Biological Productivity of the Indian Ocean, Indian J. Mar. Sci., 6, 122-137.

Redfield, A.C., B.H. Ketchum and F.A. Richards, 1963. The influence of organisms on the composition of sea water. In: The Sea. M.N. Hill (Ed.), Vol.2, Interscience, New York 26-77.

Rixen, T. and Haake, B., 1993. Fluxes and decomposition of organic matter in the western Arabian Sea. Amino acids and Hexasamines, In: Monsoon Biogeochemistry, Ittekkot, V. and Nair, R.R. (Eds.). SCOPE and UNEP, Hamburg. 113-129.

Sclater, F.R., Boyle, E. and Admond, J.M. 1976. On the marine geochemistry of Nickel, Earth and Planet Sci. Lett., 31:119-128.

Sen Gupta, R. 2001. Estuaries of Littoral Countries other than India. In: The Indian Ocean – a perspective. Sen Gupta, R. and Desa, E.(Eds.) Vol.1, Oxford and IBH, New Delhi, 299-326.

Sen Gupta, R. and A. Pylee. 1968. Specific alkalinity in the northern Indian Ocean during the south west monsoon. Bulletin of the National Institute of Sciences, India. 38, 324-333.

Sen Gupta, R., V.N. Sankaranarayanan, S.N. Desousa and S.P. Fondekar, 1976. Chemical Oceanography of the Arabian Sea: Part III. Studies on nutrient fractions and stoichiometric relationships in the northern and the eastern basins. Indian J. Mar. Sci., 5, 58-71.

Sen Gupta, R. and S.W.A. Naqvi, 1984. Chemical Oceanography of the Indian Ocean – north of the equator. Deep-Sea Research, 31, 671-706.

Sillen, L.G., 1961. The Physical Chemistry of sea water, In: Oceanography. Mary Sears (Ed.). American Association for the Advancement of Science, Washington. Pub.No.67. 549-581.

Skopintsev, B.A., 1971. Recent advances in the study of organic matter in the oceans. Oceanology, 11:775-789.

Steemann Nielsen, E and Aabey Jensen, 1957. Primary Oceanic production, In: Galathea Report, Vol.1, 49-136.

Strickland, J.D.H. and Parsons, T.R. 1968. A practical hand book of sea water analysis, Fish. Res. Board, Canada Bull., 167, 311 pp.

Strong, M. 1992. The problems and challenges of UNCED-92. Oceans and Coastal management, 18: 5–14.

Verlecar, X.N. and A.H. Parulekar, 2001. Primary Productivity. In: The Indian Ocean – a perspective. Sen Gupta, R. and Desa, E.(Eds.). Vol.2, Oxford and IBH, New Delhi, 397-415.

Williams, P.J. LeB., 1975. Biological and chemical aspects of dissolved organic materials in Sea water. In: Chemical Oceanography,

Vol.2 (2nd Edn.), Riley, J.P. and Skirrow, G., (Eds.), Academic Press, London, 301-383.

Williams, P.M. 1971. The Distribution and Cycling of organic matter in the Ocean. In: Organic compounds and the aquatic environment, Faust, S.D. and Hunter, J.V. (Eds.), Dekker, New York, 145-163.

Wooster, W.S. 1984. International studies on the Indian Ocean 1959-1965. Deep-sea Res., 31: 589-597.

Wust, G. 1964. The major deep sea expeditions and research vessels 1873-1960, Progress in Oceanography, 2: 3-52.

Wyrtki, K. 1971. Oceanographic Atlas of the International Indian Ocean Expedition, National Science Foundation, p. 58, 84..

Books for Supplementary Reading

Chester, R. 1990. Marine Geochemistry, UnwinHyman, London. 698 pp.

Duxbury, A.C. and Duxbury, A.B, 1997. An introduction to world oceans, W.M.C. Brown Pub. 504 pp.

Garrison, T. 1993. Oceanography – an introduction to Marine Science, Wadsworth Pub. Co., California, 353 pp.

Harvey, H.W., 1955. The chemistry and Fertility of sea water, Cambridge University Press, London, 224 pp.

Hood, D.W.(Ed.), 1970. Symposium on Organic matter in natural waters, Alaska University Pub.No.1, 625 pp.

Lange, R. 1969. Chemical Oceanography, Universities Forlaget, Norway, 152 pp.

Libes, S.M. 1992. An Introduction to Marine Biogeochemistry, Wiley and Sons, Inc. New York, 734 pp.

Martin, Dean, F. 1968. Marine Chemistry Vol.1: Analytical Methods; Marcel. Dekker Inc. New York, 280 pp.

Qasim, S.Z. and Roonwal, G.S. (Eds.). 1996. India' s Exclusive Economic Zone – Resources, Exploitation and Management. Omega Sci. Pub., New Delhi, 238 pp.

Qasim, S.Z.(Ed.), 1998. Glimpse of the Indian Ocean, University Press, Hyderabad, 206 pp.

Qasim S.Z.(Ed.), 1999. The Indian Ocean-images and realities, Oxford and IBH, Pub., New Delhi, 340 pp.

Qasim, S.Z., 2003. Indian estuaries, Allied Pub. New Delhi, 420 pp.

Riley, J.P. and Chester, R. 1971. Introduction to marine chemistry, Academic Press, London, 465pp.

Riley, J.P. and Skirrow, G. 1975. Chemical Oceanography, Volumes 1-3, (2nd Edn.) Academic Press, London, 606, 647 and 564 pp.

Roonwal, G.S., 1986. The Indian Ocean Exploitable Mineral and Petroleum Resources. Narosa, Pub. House, New Delhi, 198 pp.

Skiner, B.J. and Turekian, K.K. 1973. Man and the ocean, Prentice Hall, New York, 149pp.

Somayajulu, B.L.K.(Ed.), 1999. Ocean science: Trends and Future Directions, INSA and Academic Book Intl, New Delhi, 290 pp.

Stowe, K. 1983. Ocean Science, (2nd Edn.) John Wiley Pub., New York, 673 pp.

Thurmann, H.V, 1994. Introductory Oceanography (7th Edn.) Macmillan, London, 550 pp.

Turekian, K.K. 1976. Oceans, (2nd Edn.), Prentice Hall Pub., Inc., New Jersey, 120pp.

Vollenweider, R.A. 1974. A manual of methods for measuring primary production in aquatic environments, IBP Hand book, No.12, Blackwell Pub., Oxford, 255pp.

Zeitzschel, B. (Ed.), 1973. Biology of the Indian Ocean, Chapmen and Hall Ltd., London, 549 pp.

Subject Index